# The Mystery
# of the Missing
# Antimatter

*Helen R. Quinn and Yossi Nir*

ILLUSTRATIONS BY
RUTU MODAN

Princeton University Press

*Princeton & Oxford*

*Copyright © 2008 by Princeton University Press*

Published by Princeton University Press, 41 William Street,
Princeton, New Jersey 08540

In the United Kingdom: Princeton University Press, 3 Market
Place, Woodstock, Oxfordshire OX20 1SY

*All Rights Reserved*

ISBN: 978-0-691-13309-6
Library of Congress Control Number: 2007934402

British Library Cataloging-in-Publication Data is available

This book has been composed in Adobe Garamond and Caflisch

Printed on acid-free paper ∞

press.princeton.edu

Printed in the United States of America

10 9 8 7 6 5 4 3 2 1

TO TSAFRA AND DAN

# CONTENTS

# ACKNOWLEDGMENTS

We are indebted to many colleagues with whom we have worked on and discussed aspects of the science we describe in this book. We owe a particular debt to those at Stanford Linear Accelerator Center and at the Weizmann Institute and to the members of the BaBar collaboration. We have no doubt gathered tips on how to present these ideas by listening to many others do so. We have also learned by teaching; our students, be they graduate students or high school teachers in a workshop, have helped us develop good explanations by asking good questions.

This book has been a long project and in that time it has benefited greatly from the comments of several people who have read parts of the manuscript—in particular we appreciate the detailed input of John Bingham, Lesley Wolf, and Sal Glynn.

The lecture by Abraham Pais, "Paul Dirac: Aspects of his life and work," at the dedication of a plaque to Dirac in Westminster Abbey (published in *Paul Dirac, the Man and His Work*," edited by Peter Goddard, Cambridge University Press) provided much of our knowledge about the early days of Dirac's equation and much else from this early period of quantum theory. Bram's book on Einstein, *Subtle Is the Lord* (Oxford University Press) was another useful and authoritative reference for us.

The timeline section of this book began as a particle physics timeline published as chapter 10 of the book *The Charm of Strange Quarks* by R. Michael Barnett, Henri Mühry, and Helen Quinn. We are grateful for the permission of the other authors and the publishers to use this starting point (copyright AIP Press, 2000, used with kind permission of Springer Science and Business Media). The cosmology sections of this timeline owe much to web searches, and particularly to the "BrainyEncyclopedia" timeline of cosmology. The Nobel Prize website was useful in giving us biographical information on Nobel Prize winners.

# THE MYSTERY
# OF THE MISSING
# ANTIMATTER

# PRELUDE: THE MYSTERY OF THE MISSING ANTIMATTER

In the beginning—what was the beginning? Every culture asks this question. Traditionally each finds some answer, a creation myth, a cosmology. These stories satisfy an innate human longing to know about our origins. Only recently has our scientific understanding of the history of the Universe progressed to the point that we can begin to formulate a scientifically based answer—a scientific cosmology. We know that the Universe is evolving and we understand many facets of its history. We know its age, about fourteen billion years! We can ask, and often even answer, detailed questions about the very earliest times, times immediately after the Big Bang. We can test our ideas by comparing detailed observation of the Universe to detailed simulation of its evolution built on our modern understanding of physics. Today our technology for probing physics on both the tiniest and the largest imaginable scales can take us closer to the beginning of the known Universe than ever before. Much has been learned. Big questions remain; each new answer reveals new questions. What a wondrous time this is for cosmology.

Our story centers on a question that links cosmology and particle physics. Experiments in high energy physics laboratories have demonstrated that, in addition to the stuff we call *matter*, there is another set of stuff. It is just like matter except with a reversal of charges. It interacts, with itself and with matter, in ways that we understand. Physicists call this stuff *antimatter*. We make it and study it in our laboratories, but find very little of it in nature. The laws of physics for antimatter are

1

almost an exact mirror of those for matter. That makes the imbalance between matter and antimatter in the Universe a deep mystery. This mystery is the central topic of our book.

For each type of matter particle there is a matching type of antimatter particle. Given the right conditions, we can convert energy from radiation into a matched pair of newly formed matter and antimatter particles; that is how we produce antimatter in our laboratory experiments. Conversely, whenever an antimatter particle meets its matching matter twin they can both disappear, converting all their energy into a flash of radiation. Thus any antimatter particles produced in the laboratory, or in naturally occurring high energy processes, disappear again very shortly. In a matter-dominated environment their chances for longevity are very slim!

In probing the Universe today, experiments from the ground or on satellites can achieve sensitivity to times long before any structure and form evolved within it. They observe radiation that has been traveling through space for a very long time, almost as long as the Universe has existed. We can use these observations to find out about the Universe at the time this radiation began its journey. We can explore even earlier stages by modeling them according to our theories and asking whether the model can match the Universe as we observe it. The observations show a Universe that is expanding, and therefore it is cooling, and becoming, on average, less densely populated by particles.

Our theories suggest that, at very early times in the development of this evolving Universe, matter and antimatter, all possible types of particles and antiparticles, existed equally in a hot, dense, and very uniform plasma. If equal amounts of matter and antimatter had persisted, then today the Universe would be a very dull place. At the early high temperatures, creation and annihilation of energetic matter and antimatter particles would have served not only to keep their numbers equal, but also to keep those numbers large. However, as the Universe expanded and cooled, it reached a stage where annihilation could still occur whenever a particle met an antiparticle, but the reverse process, creation of a particle and an antiparticle, became more and more rare. There was essentially no radiation remaining with sufficiently high energy to cause it. Gradually all the particles and antiparticles would have disappeared. The Universe would have no visible objects in it.

Today, however, we do see a universe with huge structures made of matter: earth, solar system, galaxies, clusters of galaxies; all matter, with very little antimatter, governed on large scales chiefly by gravitational effects. All these visible parts of the structure, the stars and galaxies that light up the heavens with many forms of electromagnetic radiation, from radio waves to gamma rays (including, of course, their beautiful visible light), would not exist today if somehow matter had not won out over antimatter at some very early time in the evolution of the Universe.

How and when did the histories of matter and antimatter take such different courses? This is one of the great mysteries of science today. A question at the root of our very existence, it is one for which, as yet, science has no clear answer. Our purpose in this book is to discuss the issues around this question, explaining what we physicists do and do not understand at present, and how we hope to learn more.

We can describe the history of the very early Universe with some confidence for events that occurred from a millionth of a millionth of a second ($10^{-12}$ seconds) after the Big Bang. The reason is that in our high energy laboratories we can produce particles with energies similar to those that prevailed in the Universe at those times. We know how particles behave under those conditions. Thus, for example, the wisdom of nuclear physics allows us to model the primordial production of small nuclei from collisions starting with protons and neutrons, long before stars began to form. Because we know very well what energies are required for collisions to take apart each of the light elements into its constituents, protons and neutrons, we can identify rather precisely the time at which the Universe became cold enough that this destruction practically ceased, and thus production of elements started in earnest. This was about three minutes after the Big Bang. The success of our model of the early Universe in predicting the relative amounts of deuterium, helium, tritium, and lithium produced in this primordial nucleosynthesis is one of the great triumphs of cosmology.

For earlier events, back to as early as $10^{-40}$ seconds (a ten-billionth of a ten-billionth of a ten-billionth of a ten-billionth of a second—an unimaginably small time!) after the Big Bang, the conditions were considerably more extreme than anything that we can create in our laboratories. But, amazingly, our *theoretical* understanding of particle physics can be used to develop a reasonable (even if hypothetical) picture of events that happened at such early times, when the Universe was hot and dense beyond imagination. Our picture may not be completely right, but it builds on what we do know and tells us what issues it is critical to study further.

For yet earlier times, before about $10^{-40}$ seconds after the Big Bang, we do not even have any theory we can use; we run into contradictions if we try to apply our best current knowledge to such extreme conditions. At this time the Universe was an unimaginably hot and dense plasma of particles, interacting rapidly and energetically. The best we can do is start a moment after the Big Bang, immediately after the first rapid stage of Universal expansion (known as inflation), and follow the history of the Universe forward from then. Questions about how the Big Bang began, or what, if anything, was there before it, fascinating as they are to speculate about, cannot be addressed with any certainty at the current stage of our understanding, though they are an active topic of current work.

This much we do know: The fate of antimatter to disappear was sealed by the time the Universe was no older than a millionth of a second. At that time, matter particles and antiparticles were both still very abundant, but there must have been a tiny edge for particles over antiparticles, about one extra particle for every ten billion particle–antiparticle pairs. This tiny excess is all that matter needed for a total victory over antimatter in the present Universe. All the visible structures in the Universe that we observe today—planets, stars, galaxies, clusters of galaxies—are made from that surplus of particles over antiparticles. While we know for sure that the tiny excess of matter over antimatter existed when the Universe was a millionth of a second old, it is very likely that the crucial events that created this excess happened well before, sometime between $10^{-40}$ and $10^{-12}$ seconds, the period that is accessible to our theories but not to experiments. This makes the mystery of the missing antimatter a very exciting one: it gives us a window into extremely early times, and tests our particle physics theories under conditions that we cannot recreate in our experiments.

Matter and antimatter obey very similar but not quite identical physical laws. We know that a tiny difference between the laws of nature for matter and antimatter exists because we have seen it in experiments. It is now incorporated into our theories of particle physics. We can use these theories to develop a picture of how and when an imbalance of matter and antimatter could develop. We can even calculate how big the imbalance should be. That calculation makes predictions for conditions we can observe today, for the amount of matter in stars and galaxies compared to the amount of radiation in the background microwave signal that we see from all directions in space.

But the mystery of the missing antimatter is not solved! Our modeling tells us that the present theory of elementary particles and their interactions, the so-called *Standard Model*, which matches correctly the results of numerous laboratory experiments, must be flawed or incomplete. For, if it were the full story, the disappearing antimatter would have taken along with it too much of the matter; too few protons and neutrons would persist to make just a single galaxy, such as our own Milky Way. So this is the mystery of the missing antimatter in its modern variation: What laws of nature, not yet manifest in experiments and not part of our current Standard Model, were active in the early Universe, allowing the observed amount of matter to persist while all antimatter disappeared from the Universe?

The question of the imbalance between matter and antimatter focuses our attention on a particular early stage in the history of the Universe. Much has recently been learned about other episodes. Beautiful experiments have tested and refined our understanding of the big picture of how the Universe developed and is still developing. Along the way these experiments tell us there are two more mysteries to solve.

One is the mystery of the *dark matter*, stuff that is neither matter nor antimatter but some as yet unknown type of particle. This mysterious dark matter is also essential in the history of the Universe, but it interacts so little that it does not form stars or produce any visible emanations. We know that dark matter exists because we observe its gravitational effects. All our modeling of the motions of stars within galaxies, of the patterns of multiple images of the same distant galaxy formed by gravitational lensing as the light paths are bent by the mass of nearer galaxies, and of the evolution of structure in the Universe cannot be made to work (that, is to match the observations) without it. There is roughly six times as much mass in dark matter particles today as there is in matter particles. We cannot at present specify further what the dark matter is. All we know is that it has mass and interacts very little with itself or with matter and antimatter, except via gravitation. It is most likely some kind of particle, but we do not know enough about it yet to say what kind.

The third mystery is the most recently discovered, and hence, as yet, the least understood. The rate of expansion of the Universe is not slowing down, as we expected it should. Instead, it is accelerating. Whatever causes this is not yet part of our particle theories at all, though very speculative extensions suggest ways it might be added. This effect is generally called *dark energy*. The mysteries of dark matter and dark energy are both exciting and still developing detective stories. We will tell you what little we know about them too, though our stress in this book is on the matter–antimatter puzzle.

To develop the story of matter and antimatter in the Universe we will develop a number of themes; the Universe itself is of course one of them, along with the story of matter and antimatter. But we will also need to develop the themes of symmetries and of energy before we can tell our story in a meaningful way. Then too the theme of experiments, of how we know what we claim to know, must enter the story. We hope you will find the development of each of themes interesting in its own right; they are the basics of modern particle physics, and to understand the Universe we must understand them all.

# CONSTANT PHYSICS
# IN AN EVOLVING UNIVERSE

## Universal Laws

You cannot begin to do physics, or any form of science for that matter, without making one fundamental assumption. All science is based on it. We must assume that there are underlying laws of nature that are the same whenever and wherever we look at the world. Without this assumption there would be no predictive power to science. Whatever we might decipher of the laws of nature today in our laboratory would not be useful to predict what would happen tomorrow or in any other place. If the physical laws are independent of place and time they are universal; the laws that govern the behavior of everything in the Universe are those we can learn by experiments in any laboratory.

Of course, the laws of physics, while universal, do not say that everything is always the same. What happens in any place and time depends not only on the laws of physics but also on the physical conditions in that place at that time. The observed properties of a system of particles at very high temperature and density can be quite different from those of a dilute cold system, even though both are governed by the same underlying physical laws.

In the early days of science, the idea that the laws of science are immutable was somehow tied up with ideas about the immutability of the Universe itself. At least in the Western world, it was generally assumed that we live in a Universe that, in some overall sense, is static and unchanging, though incorporating the repeating cycles of Earth's rotation and planetary motions within that overall constancy. When Albert Einstein (1879–1955; Nobel Prize 1921) saw that his equations for general relativity (which is the name for his theory of gravity) would typically describe a Universe that evolved, he found a way to patch the theory up, adding a parameter or quantity called the cosmological constant. This could be chosen to ensure a static Universe. But we have to look at the Universe itself to decide if this solution, or one in which the Universe evolves, is the one that nature has chosen for us to live in.

The idea that the Universe itself indeed does evolve began to take shape about the same time as the idea of antimatter first appeared. These two stories, that of matter and antimatter and that of the Universe's evolution, seemed at the time to be quite unconnected. But eventually, since the Universe must work by the same physical laws as we find in our laboratory, the two stories come together in interesting ways. As we go through this book we will not follow these stories in exactly chronological order, as that would interweave the threads of the various ideas in too complicated a fashion. However, at the end of the book we provide a brief chronology so you can keep track of what happened when.

## Hubble and the Expanding Universe

The astronomer Edwin Hubble (1889–1953) was one of the first scientists to recognize that the possibility of an evolving Universe should be considered seriously. Indeed he found evidence to support it. Hubble was a brilliant man, who followed his undergraduate degree in mathematics and astronomy with a term as a Rhodes Scholar where he earned a legal degree and went on to pass the bar in Kentucky. His father apparently thought law a more respectable profession than astronomy. But eventually he returned to his original interest and completed a Ph.D. in astronomy. After World War I was over, he took up a position at the Mount Wilson Observatory from where he made enormous contributions to science.

In 1924 Hubble was the first to show that there are galaxies outside our own galaxy. But he soon went further. He noticed that the spectra of stars, for a given type of star, seemed to change with apparent brightness. He knew that, for a given type of star where the intrinsic brightness can be determined from other measurements, apparent brightness tells us about the distance to the star; like a light bulb in a long hallway, the further away it is, the dimmer it appears. As we will later explain in more detail, the data suggested that the more distant stars were apparently receding from us. The rate of this recession grows with distance.

Hubble and others argued that the pattern Hubble found meant that the Universe is expanding, which means that in the past it must have been smaller and denser than it is today. The evidence he had at the time was actually very limited by today's standards, but Hubble's inspired guess as to what it meant has stood the test of time and much better data. If you follow this argument back in time as far as you can, you come to a time, a finite time in the past, when the Universe was infinitely dense.

Other astronomers at the time, particularly Fred Hoyle (1915–2001), found this idea quite ridiculous. To replace an eternal Universe with one with a fixed starting time for its history seemed very strange to them. Indeed it raises deep questions about what was before that beginning. As yet, and perhaps for ever, these questions lie outside the realm of testable science, so we will have little to say about them. To Fred Hoyle the idea seemed outrageous enough that, to mock it, Hoyle coined the term *Big Bang* for the start of Hubble's evolving Universe. Nowadays, however, the term is used with respect, for the evidence continues to pile up on Hubble's side of the dispute. The Big Bang is the moment at which the expansion begins that leads to our currently observed Universe. We now know, with some confidence, that this occurred almost fourteen billion years ago. The Hubble Space Telescope, appropriately named after Edwin Hubble, provided important parts of the data that make us so confident of this conclusion.

The idea of an evolving Universe, together with the assumption of immutability of the underlying laws of physics, is the basis of the subject of scientific cosmology. The aim of this game is to find a model history for the Universe that emerges from the Big Bang with some simple initial conditions and that, following the known laws of physics, evolves from that instant to give the Universe as we observe it at present, a model consistent with all the observations we can make.

In this sense, the Universe can be thought of as an experimental test of our theory. It is a unique experiment! In a standard experiment, we prepare our experimental setup in some initial state, let the system evolve in time, and measure features of the final state. If we figured out the relevant laws of nature and can model their implications correctly, we should be able to predict the results of these measurements. Conversely, if we fail to do so, we may have to modify our understanding of the laws of nature.

In cosmology, the job is to run this pattern backward. The result of the "experiment" is the Universe as observed today. Can we find a simple initial state and a consistent history that would account for all the observed characteristics of the Universe? If the answer is *yes*, then the Universe experiment confirms our notions about the laws of nature and, in addition, we learn a lot about its past. If not, we must conclude that our notions are wrong or incomplete and so we must modify our theory. In either case, it is amazing how much the simple principle of requiring a consistent history can teach us, given modern observational tools and modern understanding of physics.

A suspicious reader might point out that we were not there, in the beginning, to prepare or observe the initial state. If cosmology fails to explain present-day observations, how do we know whether it is our understanding of physics that is to blame or our assumed picture of the initial state? Interestingly, many features of the Universe do not depend on details of its initial condition. But another answer lies in our concept of "a simple initial state." If we must narrowly specify too many details of that state to make things come out right, then we begin to suspect that we may not have the right physics.

This is a philosophical position, rather than a law of physics. We know that there are physical systems where the outcome does indeed depend very sensitively on the initial conditions—such as the commonly invoked butterfly effect. We explore this mystery in the hope that something as fundamental as the fact that we are here, because matter dominated over antimatter at some point in the history of the Universe, is not just such an effect. Interestingly, eventually the laws of physics will tell us whether our hope is correct, or whether in fact the only possible answer is a finely tuned initial condition. At present everything we do understand about the problem seems to suggest this is not the answer.

In this book we unravel the clues to the mystery of the missing antimatter and search for other possible answers to this question. To do this we will review the history of the Universe multiple times, with increasing attention to the details relevant to this particular mystery. This is the way scientists work. At the first pass we are happy to get the big picture right, even if some of the details are treated very roughly. Once we have the big picture we can refine those elements of the theory that do not change the overall story but do affect the details. By these successive refinements we arrive at a modern and more complete theory that incorporates all the observations and experimental results we have and suggests new tests.

New ideas seldom completely replace old ones in science, even when the new ideas are quite revolutionary. Rather we learn that the old ideas are an approximate theory that works in certain conditions, and we discover where the new theory gives the same answers and where it gives quite different ones. Einstein refined Newton's laws when he introduced the special theory of relativity. His refinement gave the same laws for conditions where they had been well tested, but an entirely different picture for the physics of objects traveling at or near the speed of light. Quantum physics was an entirely new set of ideas about the physics of atoms, but it does not change the way we think about the behavior of baseballs. We will see this pattern at work again and again as we learn what must be understood about matter and antimatter to try to understand their history in the Universe. Our newer theories almost always incorporate much of the older ones, but extend them in critical ways.

Rather than starting the story of the Universe at the beginning, we start with the Universe as it looks to us today. This is, after all, the only thing that we can observe. We then work backward to find the key points in its history for our particular question. At first we will pay little attention to matter and antimatter, and the differences between them, because this is only a very minor issue in the overall picture. We need first to understand some important general facts about the evolution of the Universe.

This understanding will allow us to run the clock backward, arriving at a simple picture of the *early Universe*. Then we will introduce matter and antimatter into our picture of the early Universe. Applying the laws that govern their interactions, we will run the clock forward again, this time following what happens to the matter and antimatter. We will arrive at a simple picture of matter and antimatter today. But it is not a picture that is at all consistent with current observations!

This is the first version of the mystery. Like good detectives we must return again and again to the scene of the crime, to the underlying laws of physics, for further evidence, and then create a second and a third scenario based on our more refined knowledge. When that does not work we must speculate further, and so we offer some ideas as to how the physics must be extended to fix the story, and a second very different scenario that could possibly resolve the problem but with a different extension of the physics. We also address how experiments can test these two speculative extensions of the theory.

## Red-Shifts: Evidence for an Expanding Universe

How do we know that the Universe is expanding? Take a drink from Alice's magic bottle to make yourself as tiny as a bacterium. Now imagine yourself standing on a raisin in the middle of a large lump of dough that is rising—that is to say, expanding (figure 2.1). No matter where in the dough you stood, you would see that all the other raisins in the dough were moving away from you, and that they were moving away faster the further they were from you. We see this same pattern for the objects in the sky. We look at stars in distant galaxies, as far out as we can see them. They seem to be moving away from us, and the further out they are the faster they appear to recede.

You might think this means we must be at the center of the Universe. Not so. Go back to the dough. Whichever raisin you sit on, all the others appear to be moving away and yours seems to be still (unless you are at the edge and can see out, but the Universe has no edge).

How do we know that the stars are moving away from us? This is an inference we make by applying well-understood physical laws to understand the patterns of a star's light spectrum compared to the distance to the star, as Hubble first did. Any star emits light missing certain characteristic frequencies. These frequencies are those absorbed by atomic transitions in the hot gases at the surface of the star. Photons streaming toward us are absorbed, causing an electron in an atom to jump to an excited state. The reverse process, photon emission when the electron falls back to the ground state, occurs, but the photons go in all directions, not just toward us, so we see fewer photons at the particular frequencies that atoms in the gas can absorb. Each type of atom has its own set of absorption lines, but every star is surrounded by gases that include certain common

Fig. 2.1 Imagine yourself standing on a raisin in the middle of a large lump of dough that is rising.

elements, such as hydrogen or helium. So the absorption lines of those elements are a very useful key to use in deciphering the motions of distant objects.

We slipped some new words in here—photons, excited states—and with them come a lot of important physics ideas, so it is worth taking a moment to be sure that these ideas are at least a little familiar to you. We will develop them at greater length later in our story. A photon is a massless particle. Light can be viewed either as a traveling electromagnetic wave, or as a stream of these particles. Actually it has properties that we associate with waves and others that we associate with particles. This duality is one of the weirdnesses at the core of all quantum physics—all particles sometimes behave like waves, and likewise light, classically viewed as a wave phenomenon, can be seen as particle-like as well. Einstein's 1921 Nobel Prize was awarded for his 1905 work showing that this particle property of light was necessary to explain the photoelectric effect—absorption of light causing emission of an electron. The absorption lines seen in stellar spectra can be understood only with a quantum view both of atoms and of light—indeed the explanation of atomic absorption or emission lines was a key early success of quantum theory.

The idea of an excited state of an electron in an atom is also a quantum concept. In quantum physics we find that the electrons in an atom can only be found with certain definite energies, and with configurations that are unique to those particular energies—where by "configuration" we mean the probability distribution for finding the electron anywhere around the nucleus, and the angular momentum of the electron relative to the nucleus. There are a calculable set of allowed quantum states for electrons in any atom, and since each one can be occupied by only one electron, their patterns define key features of chemistry. The lowest energy state possible is called the ground state for that atom, and an excited state is one where one or more of the electrons are in a configuration that has higher energy than its lowest possible state. Because there are only certain discrete energy states, there are thus certain definite photon energies that can be absorbed; the photon disappears and kicks an electron up to a higher energy state. If the photon energy is great enough to kick the electron right out of the atom we have a continuous absorption spectrum, but there are a discrete set of lower energy absorption lines that are characteristic for each type of atom.

Thus the pattern of these dark lines, the absorption frequencies, for a star at rest relative to us, depends only on having particular types of atoms present in the gas around the star. We can reproduce the pattern for a given type of atom by heating a gas containing those atoms in the laboratory and looking at the spectrum of radiation from it, because any atom that can absorb a particular frequency will also give off radiation at that frequency when it is hot enough. However, we find that the pattern we see for each star matches that we see in the lab only after we make a correction for the shift in frequencies due to the motion, or apparent motion, of the star relative to us.

The effect of relative motion of the star and the observer is called the Doppler effect. It is familiar to anyone who ever heard the motor of a racing car going by at high speed. As the car moves toward you, you hear a higher frequency (pitch) than for a stationary car with its motor running, and as it moves away you hear a lower pitch. The motor does not change, but what we hear does. We can understand this when we recognize that sound, and light, are waves. The pitch of the sound, or the color of the light, depends on the frequency with which the successive peaks in the wave arrive at our detector (ear or eye). If the source is moving toward us, the peaks in the wave it emits get crowded together between it and us, and so we hear (or see) a higher frequency but, as it moves away from us, the peaks get spread and we detect a lower frequency. The spectra of the stars can be explained if they are moving relative to us. Every absorption line can be moved to its standard frequency by the same choice for the speed of the star relative to us.

There is a second effect that looks to us just like a Doppler effect, but is in fact quite different. Distant stars appear to us to be moving away, but, like the raisins in the dough, they are actually not in motion with respect to the part of space in which they exist. Instead the frequency of their emitted light has been lowered by the expansion of the universe during the time that the light is traveling from them to us. This expansion stretches out the wavelength of the light and hence lowers its frequency. If we interpret this shift as a Doppler effect then very distant stars appear to be moving away at speeds greater than the speed of light. This is not forbidden, because this is not an actual motion of the galaxies, a speed relative to the space-time in which they exist, but only an apparent speed due to the universal expansion. Think again of the raisins in the dough; they are moving apart, but not one of them is moving relative to the dough immediately around it.

The shift of spectral lines due to the expansion of the universe is called a *red-shift* because lowering of the frequency of light within the visible spectrum moves any light toward the red end of the spectrum. However, this shift can be so large that lines are moved from the visible region into the microwave, while other higher-frequency lines are shifted from outside the visible range to any visible hues. So red-shifted stars do not necessarily look particularly red. Indeed, many distant galaxies look quite blue to the eye, because their higher-frequency X-ray radiation has been shifted into this visible region by the expansion of space since the radiation left them.

The first indication that this idea is correct is that, for any given object, there is a single choice of red-shift that assigns shifts so that *all* the spectral lines move to their natural values—the ones we measure in the lab. This tells us that we have correctly interpreted the shifts as being due to either the motion of the star or the expansion of the universe. (Or some of each. Consider the stars in a rotating galaxy: due to the rotation some are moving toward us, some away. If, at the same time, the entire rotating galaxy appears to recede because of the expansion of the universe then all its stars may appear to be moving away, but some a little faster and some a little slower than the average.) From the spectral shift of a single object alone, say a single supernova in a remote galaxy, we cannot tell whether we are seeing the effect of a Doppler shift due to real relative motion of the galaxy drifting away from us through space, or a red-shift due to the expansion of the intervening space, but when we look at many galaxies we can disentangle the patterns.

Now the important part of the result is that when we calculate these motions we find that the more distant a star is from us the larger, on average, is its red-shift. This effect is so overwhelming that the small corrections due to local motion do not wipe out the overall pattern of red-shifts growing with the distance to the objects.

How we know the distance to a given star or galaxy is another interesting and complex story, particularly for the most distant objects, but one that we will not discuss in detail here. The basic idea is this: The brightness you observe for a light of any standard type of source decreases as you increase the distance between yourself and it. Anyone who has driven a car at night knows this.

So, if we can observe standard types of stars, that is, stars for which we know their characteristic intrinsic brightness, we can tell how far away they are by how bright they appear to be. It turns out that there are indeed certain types of stars that provide such standard brightness sources. This is sufficient to establish a rough law that red-shift grows more or less linearly with distance. The only reasonable source of such a law is an expanding Universe. Hubble's first example of a standard star was a type of star called a Cepheid variable; more recent observations use a particular type of supernova, which, with modern methods, can be detected over much greater distances.

The fact that the universe is expanding was first proposed by Hubble in 1928 based on his observations of red-shifts. Recent evidence from observations of distant supernovae dramatically confirms this basic picture, but extends it to tell us also that the rate of expansion is changing in a quite unexpected way. The picture of an expanding Universe predicts a number of other effects, as we will explain below. Its most important predictions have been confirmed in some detail by very precise measurements made in the past few years. The new discovery, on the way the rate is changing, opens new questions at a next level of detail, but does not change these overall features.

Ideas about cosmology that originally seemed to be untestable speculations have recently been remarkably well tested. Cosmologists now have a standard model too; they call it the "concordance cosmology." It is a model of the properties of the Universe that fits the data from a great variety of different types of observations. The supernova red-shifts mentioned above are just one of these. Another important observation is in fact the oldest thing that we can directly observe. It is radiation that arose in very early times in the history of the Universe, signals that have been traveling across space for nearly fourteen billion years, essentially unaltered during their journey, except by the red-shift from the expansion of the Universe. Today we see it as microwave radiation coming to us from all directions in the sky.

Detailed observations made of these signals in recent years, by ground and by satellite experiments, from COBE (launched in 1989) to WMAP (launched in 2001), give us detailed information on the state of the very early Universe.

So now let us run the clock backward, shrinking away that expansion. The earlier we go, the smaller any region of the Universe becomes. If the Universe is filled with a gas of particles, then at early times it was denser and hotter than it is today. This is roughly all that we need to know to start running the clock forward again and follow matter and antimatter: at some very early time, the currently observable Universe was very small, very dense, and very hot.


## Numbers Large and Small

As you have already seen in the previous chapter, to talk about the history of the universe, we need to comprehend some incredibly large numbers, and also some extremely small ones. Likewise, in the world of particles we must deal with extreme scales, tiny sizes and very short lifetimes. One of the beautiful things about this story is how the science of the very large and that of the very small are interlinked; we cannot understand the first without understanding the second. Already we began to use the scientific shorthand for large and small numbers, because it is just too clumsy to write them out in their everyday form. We must deal with too many powers of ten. It is easy enough to write and read 1,000 rather than $10^3$, or 1/100 rather than $10^{-2}$. But when we get to numbers as large as $10^{30}$ (the mass of the sun in kilograms) it is too cumbersome to write 30 zeros after the 1, and when we get to numbers as small as $10^{-30}$ (the mass of the electron in kilograms) it is equally hard to count all the zeros in the denominator.

The code is easy enough; the exponent after the ten tells us how many powers of ten, how many zeros we would have to write after the 1. If it comes with a minus sign then we must divide by that many powers of ten. What is hard is to comprehend these numbers—so large or so small. It is almost impossible to imagine what they mean. Indeed it is fair to say that even the scientists who use them all the time have no real conception of how big or how small a quantity is represented by numbers like $10^{26}$, the ratio between the size of the observable Universe and height of a typical six-year-old child, or $10^{-18}$, the ratio between the smallest distance probed in our accelerators and the same child's height. We just get used to thinking in terms of exponents. We do recognize that each factor of ten takes us to a quite different scale, and are awed by the fact that we understand enough of the laws of physics today to range freely across so many scales in this story.

Of course we can only do this because we ignore a lot of very important details that occur at intermediate scales. In this story everything between an atom and a cluster of galaxies is pretty much a detail that we ignore. That does not mean that we think these intermediate scales unimportant or uninteresting—all of chemistry, biology, earth science, and astronomy, not to mention most of what physicists and astrophysicists study, lies in these intermediate scales. We ignore them to paint the big picture of the Universe and its history. But we cannot also ignore the details of the very small scale physics and get that picture anywhere near right—the high-energy particles that populate the early Universe make that impossible.

## What Do We Mean by "Universe"?

You may have heard it said that the Universe was infinitely small at the time of the Big Bang. This would be true if the Universe were finite in extent today, but not if, as current evidence suggests, it is infinite. If the Universe is infinite in extent today it was also infinite in extent then, as well as infinitely dense. Not an easy thing to think about. However, the difference is somewhat semantic, as the entire *observable* Universe, by which we mean that part of it from which Earth can ever receive any signal, is finite in extent in either picture, and it grew out of a region that was very, very tiny at the time of the Big Bang.

Astronomers tend to use the terms "the Universe" and "the observable Universe" interchangeably; that is the only Universe relevant to their observations. Cosmologists may make the logical distinction, because their theories may describe not only the part of the system that we can observe, but also parts that are outside any possible contact with us. When cosmologists talk about the Universe, you have to ask whether they mean our observable Universe or a much bigger system of which our observable Universe is just a tiny part. Often they are talking of the much larger system.

Perhaps this is a point where our knowledge has evolved to the stage where old definitions of the word Universe no longer work and we need to define new words. In cosmology today the word "multiverse" is sometimes used to describe a system that contains many disconnected regions that will never have any communication with one another. In this language, our own Universe is perhaps just a small part of one of many regions. Of course, we will never be able test any theory about what is beyond our observable Universe, except insofar as a theory that includes a multiverse does or does not give a consistent history for the part of the system that we do observe.

In this picture, the Big Bang is not a true beginning, but simply the time at which our particular region of the multiverse began an exponentially rapid period of expansion, known as inflation. All the details that we talk about in this book begin once that inflationary period is concluded and the Universe emerges from it, filled with a hot dense gas of particles and radiation, and beginning a history of a much slower pattern of expansion (in which space grows over time with a power-law relationship, rather than an exponential relationship between the size of any region in the space and the time elapsed). Whatever occurred before that time, or whatever exists outside our visible Universe, has no impact on our story and so we will not discuss it further. Indeed it has no impact on anything we can ever measure, so we can never do more than speculate about it, and thus, interesting though it is to think about it, it is outside the realm of science.

When we run the clock forward in the following sections, following either the radiation or the matter and antimatter, we will do so from *almost* the beginning of the Universe. What happened "in the beginning" is still a challenge. The conditions immediately after the Big Bang were so extreme, with such high density (above $10^{90}$ gram per centimeter cubed) and temperature (above $10^{32}$ kelvin) of matter, that we don't even know the laws of physics for those conditions. Indeed, each particle collision is so energetic that we know that gravity is as strong an effect in such a collision as any other force, and our current theories of matter do not allow us to treat such strong gravity in a consistent way. Hence we cannot really follow the story all the way back to any definite start.

However, that doesn't much matter: Whatever the initial situation, a tiny fraction of a second later the hot dense plasma of particles had expanded and cooled to the point that both the density and the average particle energy were such that we can begin to probe similar conditions in laboratory experiments. Since it was hot and dense we can use our knowledge of thermodynamics to determine the general properties, because in any such system thermal equilibrium is very rapidly achieved. Thermal equilibrium simply means that, because collision processes redistribute energy among the particles and radiation very rapidly, they all share a uniform temperature; the energy of the system is shared evenly among them, just as it is for multiple different gases in the same container—you cannot have a mixture of two gases at different temperatures—collisions share the energy between them and bring them to a common temperature. So we begin with a hot, dense, thermally uniform plasma of particles and radiation, just after the Big Bang. And it is here that the mystery of the missing antimatter begins.

# 3 AS THE UNIVERSE EXPANDS

## Running the Clock Forward: Radiation

Let us set our clock to zero at the time of the Big Bang and follow the story of the Universe, this time running our clock forward and watching some tiny region of the Universe as it expands. But before we get into details about matter and antimatter, we need the overall picture so, to start with, we will run the clock forward focusing on what happens to radiation and particles, but not distinguishing between matter-particles and antimatter-particles for now. We will summarize the evidence that the Universe is indeed expanding and that Hubble and those who followed him got the basic story right. Recent experiments have tested these ideas in some detail. Putting together all the evidence from a variety of different probes gives a remarkably precise overall picture of the general properties of the Universe. This is called the concordance model of the Universe.

The way the game of cosmology works is that, once we have any clue from data, we make a model for the history of the Universe that has this property. Then we ask what else that model predicts. If we then look for that effect and find it does occur, our model is on the right track. If not, we have to throw out that model and start again. We cannot go to the laboratory to test these ideas; we must use the Universe itself as our laboratory. Our clue here is that the red-shifts of different stars suggest that the Universe at early times was much hotter and more densely populated with particles than it is today. But this is no ordinary gas expanding into a previously empty space, this is space itself expanding, diluting the density of diffuse matter within it. Even so, many of the familiar features of expansion apply, though for somewhat different reasons.

Any gas of particles and/or electromagnetic radiation gets less dense as the universe expands; there is more space for the same stuff. At the same time, in an expanding Universe, the radiation and particles present get stretched to longer wavelengths. Now, for both massive particles and radiation, longer wavelength means lower energy. For massive particles the lowest possible energy is the mass energy, $E = mc^2$, the energy of the particle at rest. Once the wavelength gets long enough or, in particle language, once the motion is slow enough that the mass energy is what dominates, the energy per particle is more or less constant as the Universe expands. However, for photons (or any other massless particle) the energy per particle keeps getting lower and lower as the wavelength stretches longer and longer.

The temperature of a gas is a measure of the average motion energy per particle or photon in the gas. For any gas in thermal equilibrium, there is a characteristic distribution of motion energy. A graph of the probability for finding a particle at each energy has a shape that depends on the temperature, with the peak probability moving to higher energy with higher temperature. Furthermore, if there is radiation interacting with the gas, it too has a characteristic distribution of energies, which means it has a characteristic spectrum, known as a thermal or black-body spectrum. The form of this spectrum is very well known and understood.

Indeed quantum physics, which is the basis of all our modern particle theories, began in 1900 with the explanation of this spectrum. Max Planck (1858–1947; Nobel Prize 1918), a professor at the University of Berlin, presented two papers before the German Physical Society, one in October and the second in December. In the first he presented a formula for the spectrum of radiation which fitted the data with a few parameters to be chosen arbitrarily. His derivation was incomplete, but the formula clearly worked; it fitted the data as no previous attempt had done. Planck's second paper showed that he could derive his formula and thus explain the observed spectrum, but only by assuming that radiation had a quantum property, that it comes in definite amounts, which we call quanta. He had to assume that each quantum must have a definite energy ($h$ times its frequency), where $h$ is a defined constant quantity. Planck could use the known spectrum to determine what the value of the quantity $h$ must be. This fundamental constant of nature is now known as Planck's constant. Once he assumed this rather surprising property for the radiation he could give a derivation that reproduced his earlier formula, with the unknown quantities replaced by $h$.

This was revolutionary! The only allowed energy for the amount of radiation (produced in interaction with a hot container) at frequency $f$ is any whole number of quanta of that frequency. Until then, anyone you asked would have said you could have any amount of radiation with frequency $f$ that you wanted. Planck's bold assertion that you could not, that radiation comes only in quanta of ($h$ times $f$), or at least that the radiation coming from a hot black object can only have these particular energies, was the first step in the theory of quantum processes. It seemed quite weird, but it had the advantage that it then gave the observed spectrum. All previous attempts had failed to explain this spectrum. Now we know that this is a property of photons, quite independent of how they are produced, and we have a mathematical theory of photons and their interactions with matter that gives these quantum properties.

Later, physicists realized that we can assign massive particles wave-like properties similar to photons and thus they too have a definable wavelength or frequency. Here, however, one is not surprised by the fact that one has only integer numbers of particles, but rather by the fact that they behave like waves, as does light. That is why we talked about wavelength for massive particles earlier. (We are getting ahead of ourselves in the particle side of the story; these quantum ideas deserve more explanation and we will get to that later.)

So each particle or photon (the quanta, or particles if you like, of electromagnetic radiation) has an energy that depends on its frequency or wavelength. As the Universe expands, all wavelengths get stretched uniformly. If at any time the particles and radiation in the Universe have a spectrum that is a characteristic thermal distribution for a given temperature, then, at a later time, the particles and radiation will still have a characteristic thermal spectrum, but it will correspond to a lower temperature. The Universe automatically cools because space expands. Modern observations of the microwave radiation that fills the Universe show it has the expected Planck spectrum with incredible precision, telling us that the Universe is the most perfect thermal system that has ever been observed.

Now consider a Universe filled with particles and radiation at a very high density and temperature just after the Big Bang. How would it evolve? What should we expect to see today if this were the correct picture for the start?

In a very hot gas, the average energy of the collisions between atoms is high enough to knock electrons off atoms, and thus turn them into ions. Make the gas hot enough and it becomes plasma; a gas of charged particles, electrons and ions. Such plasma is not transparent to light, because photons interact with charged particles. Hence they are scattered as in a dense fog, rather than traveling straight ahead. Charged particles also radiate photons when they collide with one another, so in the plasma there are always photons present. Because they interact frequently, the photons will be in thermal equilibrium with the ions and electrons, and the whole system can be characterized by a single temperature. Make it even hotter and the collisions begin to knock apart the nuclei. Eventually, at high enough temperature we would get a gas that contained protons, neutrons, electrons, and photons, all moving around rapidly and colliding frequently. So we know this was how it was when the Universe was hot enough.

Now reverse the process; let the system expand and cool. As the temperature drops, so does the average collision energy. Nuclear fusion processes can form light nuclei. Eventually the gas cools enough that when nuclei form they are not knocked apart again in collisions. So then we have a gas with a mixture of nuclei, as well as electrons and photons. Eventually we reach the point where the average collisions no longer cause ionization of atoms. They do not have enough energy to do so. Then gradually the ions and electrons meet and combine to form atoms. The plasma becomes a gas of neutral atoms and is transparent to light. As the Universe expands and cools it too must go through these same stages or transitions. Each of them leaves important watermarks in the current Universe, evidence we can use to tell us about its earlier state.

Once the Universe cools to the point where a hydrogen atom is stable in the typical collision, a temperature of about 3,000 kelvin (that is, centigrade-sized degrees above absolute zero temperature), the electromagnetic radiation present at that time (loosely referred to above as light) no longer finds any charged particles to interact with. Instead it begins its long journey across space and just keeps going. As it travels it is gradually red-shifting, and thus cooling, due to the expansion of the Universe. For over thirteen billion years it has been traveling through space and getting colder and a longer wavelength. Eventually it became microwave radiation. Its spectrum today corresponds to the much lower temperature of 2.73 kelvin. We detect this radiation, a relic of the time that the Universe first became transparent to light. That happened when the Universe was about three hundred thousand years old. When Arno Penzias and Robert Wilson first noticed a pervasive microwave signal in their radio telescope they thought it was an instrument effect and tried very hard to get rid of it. But conversations with their cosmologist friends at Princeton soon led them to realize it was a real signal from space; they were seeing light from the beginning of the Universe. They were awarded the Nobel Prize in 1978 for this discovery. More recently this "cosmic microwave background radiation" has been studied in great detail, thereby providing interesting information about the very early history of the Universe and remarkable tests of some major features of this story.

In whichever direction we look out into space today, we see this radiation arriving. The amazing observation is that, once we correct for the Doppler shift corresponding to motion of the Earth relative to this radiation, we see a precise thermal spectrum at the *same* temperature, to within a few parts in a million, no matter which way we look.

The Nobel Prize in 2006 went to John Mather and George Smoot for their leadership of the COBE satellite project which made the first modern and really precise measurement of the spectrum and its temperature, and found the tiny temperature variations splattered across the sky. This precise thermal spectrum is the best evidence we have for the expansion of the Universe. This radiation is the oldest thing we can ever directly detect. The patterns of variation in its temperature, tiny as they are, contain rich information about the history of the Universe before this radiation began its many-billion-year journey across space.

This picture of Universal expansion gives the only reasonable explanation of the remarkably accurate thermal distribution of energies in this radiation and its uniformity of temperature. In fact, it is so uniform in temperature that it very strictly limits the amount of variation in density that could have been present at the time this light began its nearly fourteen-billion-year journey through space. All the structure we see now, the galaxies and clusters of galaxies, evolved by gravitational collapse from tiny fluctuations in the density of that hot plasma that produced this light. These density variations leave their imprint on the radiation as tiny temperature variations. So the pattern of variations in background microwave temperature probes the patterns of that protostructure. This pattern is now very well measured. It tells us how to start our simulations for growth of structure in the Universe, and furthermore it provides a test of theories about the very earliest times. It supports the idea that there was an early epoch of inflation (a period of exponentially rapid expansion at a much earlier time) because the theory of inflation predicts certain major features of this pattern correctly.

For all practical purposes, the story we are investigating here begins when the rapid inflation ends and the Universe settled into a type of expansion more like that we observe today.

## Running the Clock Forward: Dark Matter

In the next stage of the history, after the atoms form and the Universe is transparent to light, all the structure we observe in the Universe today must evolve from the protostructure, the small density fluctuations that cause the temperature fluctuations observed in the cosmic background radiation. These fluctuations are tiny; the temperature varies only by a few parts in a million over the entire sky (if we look at it from the right frame of reference). However they are critical.

It takes large computer models to simulate what happens. Because gravity is an attractive force, any region with a small density excess tends to contract and become even denser. But there is a competing effect; the motion of particles tends to smooth out any variations in density. In an ordinary gas the latter effect wins and gravitational effects are tiny, but in the history of the Universe it is the gravitational effects that dominate the picture. Starting from tiny variations in density that developed during the initial inflationary period, gravity has formed all the structures we see: stars, galaxies, and clusters of galaxies, all made from matter.

But most of the mass in the Universe turns out to be neither matter, nor antimatter, but some other kind of stuff that we call dark matter.

Most of the mass in any cluster of galaxies is not the matter which forms stars (and planets) but some other, much less interactive, type of particle, and this is what is known as dark matter. The dark matter forms a broad and relatively smoothly varying mass distribution throughout a cluster of galaxies, within which the galaxies are small bright regions, which correspond to sharp peaks in the density distribution.

Dark matter was first suggested based on two types of observation that are sensitive to the presence of this dark mass. The first reason to suggest its existence, observations that showed that there is dark matter within galaxies, was to notice the effect the mass of the dark matter has on the motion of the stars about the center of mass of the galaxy. The motion of outer stars in spiral galaxies could not be explained by the mass of the stars inside their orbits; there had to be much more mass than that to provide the gravitational pull causing the orbital motion of these stars and holding the galaxies together. A second, more recent, set of observations sees the gravitational effect of the dark matter in one galaxy on the light coming from more distant galaxies behind them. This "gravitational lensing" effect produces multiple distorted images of the same distant galaxy. The information in those distorted images can be used to decode the map of the distribution of mass that produced them. It is found that the mass of any galaxy extends well beyond the region containing stars.

The ordinary matter clusters into even denser regions than the dark matter because it is interactive and thus experiences collisions which provide a dissipative mechanism to damp out large individual particle momenta, redistributing it among many particles of lower momenta. Thus ordinary interacting matter becomes more clumped with the same gravitational pull than the dark matter, which does not experience any such collisions. Once ordinary matter gets dense enough, stars form and begin to shine due to nuclear processes deep in their cores, just as does our sun. Because we see the stars shine, we can tell that they contain ordinary matter, with its known interactions. Indeed we can use the physics we understand from laboratory experiments to explain in great detail how stars shine.

We also have a third argument that points to the existence of dark matter, and even tells us how much of this stuff there must be in our Universe. Using computer simulations astrophysicists explore how the observed pattern of galaxies and clusters of galaxies could be formed starting from the density fluctuations mapped by the temperature fluctuations of the cosmic microwave background radiation. They find that they simply cannot get to the observed patterns of galaxy clusters without including dark matter in the simulation. If one puts all the mass density into ordinary matter, the models give a Universe that looks completely different from the one we observe; it does not have the right patterns of galaxies and galaxy clusters. Similarly, if one puts in too much dark matter the patterns are also quite different. Furthermore, since we know from the other two observations that there is dark matter, we know we must include it, and require that the models arrange it so as to give the gravitational lensing and star motion patterns that we observe.

The ratio of dark matter to ordinary matter is thus pinned down quite precisely by the requirement of a concordance model that fits many quite different experiments. These experiments also tell us something else is there, something we know even less about, to which we give the name dark energy. The data include the patterns in the cosmic microwave background radiation and in the structure of the Universe today, the red-shifts of distant supernovae, the patterns of absorption lines due to hydrogen in interstellar gas clouds, and the accounting of mass density in stars. These come together to tell us that our Universe is a flat space, and this property tells us the total energy density. Furthermore, the combined experiments tell us that about 70 percent of this energy density is the mysterious thing we call dark energy, an effect that is beginning to speed up the expansion of the Universe, as first shown by the observation of distant supernova explosions. This was a big surprise; scientists had all expected these observations to show that the expansion of the Universe was beginning to slow down as gravity pulled the expanding Universe back together. Instead they showed it speeding up. Such an effect requires something quite new, not just a new kind of particle but something we understand very little about, to be added to the mix. But even though it dominates the energy density of the Universe today, it has only recently come to do so, as expansion dilutes the other forms of energy but not this one. So in fact dark energy plays almost no role in our story; it simply stands as a new question mark at its end.

The remainder of the energy density is now in mass, the mass of some kind of particles. Of the particles, nearly six times as much of the mass is in the dark matter particles as in ordinary matter particles. This fits the data from gravitational lensing and those from the motions of stars in spiral galaxies, as well as the evolution of structure. In the early Universe a significant fraction of the energy was radiation, but that has now been red-shifted to such low frequencies that it contributes little to the total energy budget, even though we do see it as the microwave radiation coming from all directions.

We do not yet know what the dark matter is. We know it contains particles that have mass, and have very little interaction either with one another or with the particles that we are made from, other than gravitational effects. The dark matter is certainly not antimatter, since that would behave more like matter. Antimatter is just as self-interactive as matter, and so it too would condense into extradense regions and eventually make stars. It would not be spread out throughout a galactic cluster as the dark matter is observed to be. Furthermore, the known interactions between matter and antimatter would create dramatic effects that we do not see. So that possibility is excluded. We, particle physicists, have some ideas about what dark matter particles might be, but as yet no evidence for which, if any, of these ideas are correct. We will return to these ideas at a later point. As far as the dark energy goes, we have much less certainty what that is, though we can say that this recently discovered phenomenon is not yet another kind of matter. It is something quite different, something yet to be explained.

## Running the Clock Forward: Light Nuclei

Another set of observations that cosmology must explain is measurements of the relative abundances of various stable light element nuclei (hydrogen, deuterium, helium, and lithium) in the primordial gaseous regions of the Universe. Primordial here means the regions with no stars, because stars also make nuclei by fusion processes inside their cores. If we want to know about the earliest nuclei formed by fusion as the Universe cooled just enough for nuclear stability, we must look in places not "polluted" by stuff from stars (or as Carl Sagan called it "star stuff"). The data from these regions provide a window to an earlier stage in the history of the Universe, before it became transparent to light. It tells us about the time when the average particle energies first cooled to the point that nuclei were stable in the average collisions. This occurred a few minutes after the Big Bang. This chapter in cosmology is called "primordial nucleosynthesis." The impressive success of cosmology in this regard makes us confident that the Big Bang theory puts us on the right track. But, for our detective story, there is an extra bonus in the study of nucleosynthesis: it tells us quite precisely how much ordinary matter there is in the Universe compared to radiation. That ratio is something that must be explained by any good theory of how matter won out over antimatter.

We observe stars, galaxies, clusters of galaxies, and even superclusters. All this structure takes place on rather small distance scales compared to the observable size of the Universe. On very large scales, the Universe is seen to be, to a large extent, empty. The average density of matter within a galaxy is about a hundred thousand times larger than the average density of matter in the Universe. Direct observations of matter in the Universe allow us to set an upper limit on the overall amount of matter; there is not more than one proton or neutron for every hundred million photons in the Universe. Most of these photons are in the all-pervasive microwave radiation we just talked about; all the activity coming from stars is just a small addition to the overall number of photons in this relic of the hot early Universe.

But when you set an upper limit on a quantity (it is not more than one proton or neutron for every hundred million photons) the quantity could actually be much smaller than that limit. This is where primordial nucleosynthesis plays a role in our story: it pins down the number we want, giving us a quite precise test of whether our matter–antimatter story is right.

There are two key quantities that determine the ratios of different types of nuclei formed in the primordial nucleosynthesis. The first is the ratio between the number of neutrons and the number of protons at the time when nuclei became stable, and the second is the ratio of matter particles to photons. Given the Standard Model theory of particle physics and the concordance picture of cosmology, we know everything that we need in order to calculate the first ratio. It turns out that there were about seven times more protons than neutrons by the time the Universe had cooled sufficiently for the light nuclei to be stable in the average collisions. This difference occurs both because neutrons are somewhat more massive than protons, which means that in thermal equilibrium there are fewer of them, and also because free neutrons are unstable, and hence the ratio changes with time as neutrons decay. So the ratio depends on the time it takes for the Universe to reach the temperature at which light nuclei are stable in collisions. Most of the neutrons present at this time end up in helium nuclei, each of which consists of two neutrons and two protons. This fact allows us to predict in a very simple way the ratio between primordial helium nuclei and primordial hydrogen nuclei (that is, protons): every two neutrons will combine with two protons to make a helium nucleus. Since there are seven times more protons than neutrons, for every two protons that end up in a helium nucleus, there remain about twelve more that are not able to find neutron partners to make helium nuclei. Thus, the ratio of helium to hydrogen should be about one to twelve, which is in beautiful agreement with observations.

In addition to the nuclei of hydrogen and helium, there are a few other light, stable nuclei that are produced primordially, though at much smaller numbers. To make detailed predictions on their relative abundances, we use our knowledge of nuclear physics to model all the processes of nucleosynthesis. Fusion builds simple nuclei from collisions of protons and neutrons with one another, and with the nuclei already formed. Fusion forms deuterium, tritium, and lithium, as well as helium, in the early Universe. For each type of nucleus, the number formed compared to the residual number of protons (which later become the nuclei of hydrogen atoms) can be predicted. The detailed results depend on the ratio between the number density of matter particles and that of radiation (photons) at this time. This ratio affects the temperature at which a given type of light nucleus, once formed, is unlikely to be destroyed in any subsequent collision with a photon. The expected abundances of light nuclei are quite sensitive to it. The fact that we can fit the observed ratios to protons for all four types of primordial light nuclei is another strong piece of evidence that our model, an expanding Universe, is right. But the fit is successful only for a very narrow range of the ratio of matter particles, protons and neutrons, to photons: the predicted abundances of deuterium, tritium, and lithium are very sensitive to this ratio, so measurements of the relative abundances can pin it down very accurately. The matter to photon ratio found in this way is even smaller than the limit we set by direct observation. There is a single proton or neutron for every billion photons.

So now we know from two quite separate chains of evidence that matter is a bit player in the overall history of the Universe, though of course it is central to our existence. We know that there are many more photons, albeit very low-energy ones, than there are matter particles of any kind, and we know that there is much more mass in dark matter particles than in ordinary matter particles. (What this means in terms of numbers depends on the mass of an individual dark matter particle compared to the mass of a proton, and we do not know that yet, so we cannot yet translate mass density to number density for these particles.)

## Running the Clock Forward: Matter and Antimatter

So far, we have run the clock forward three times: First, following the history of radiation; second, following the history of structure in the Universe; third, following the abundances of the light elements. In all three aspects, we can get a universe that looks like the one we actually see, but only if we begin with the right mix of radiation, particles, and dark matter. We will now run the clock forward for the fourth time, following the history of matter and antimatter: will we be able to get what we need from this—a Universe with very little matter and with practically no antimatter?

We have already explained, in the previous section, what we mean when we say "very little matter": There is a single baryon (a proton or a neutron) for every billion photons. Before we run the clock, we would like to explain more concretely also what we mean by "practically no antimatter." The central mystery for this book is that, while there are very few baryons for every photon, antibaryons are even more rare. (The word baryon here is introduced as the generic term for a proton or neutron; antibaryon for their antiparticles. In fact there are other types of baryons too, heavier and unstable cousins of the proton or neutron, but these are not present in any significant numbers at this point in the history of the Universe and so need not concern us here.) To the best of our understanding, all the visible structures are made from matter, none are antimatter. The population of antibaryons in the Universe is effectively zero. Even between galaxies, space is a lively place, continuously flooded by particles produced by the active galaxies. If some galaxies were made of antimatter, they would be radiating antiparticles. We would see the interactions between these antiparticle cosmic rays and the particle ones from matter galaxies in the regions between galaxies. Arguments like this tell us quite reliably that, out to about one-third the size of the visible Universe, all the stars are made from our type of matter. The notion that a hot dense mixture sorted itself out into a few such huge regions of matter and other, equally huge regions of antimatter is completely unphysical (no model based on known physical laws works that way; you would have to have very different laws of physics for matter and antimatter to unmix the mixture). Thus we can safely argue from that observation that our entire Universe contains only matter clusters, with none made from antimatter. The only antimatter in space is that produced in high-energy collisions, in combination with matching matter particles. Because there is more matter than antimatter, any antimatter that is produced in this way does not last long; it meets matter and annihilates, and both the antimatter particle and a matching matter particle disappear, their energy transformed into radiation.

This is then the present picture of matter and antimatter in the Universe: one baryon for every billion photons and essentially no antibaryons. But how did it get to be that way? Our starting point, at almost the beginning, was a very small Universe, filled with plasma of matter, antimatter, dark matter, and radiation, with very high density and very high temperature. Our theory further suggests that, at that stage, the amounts of matter and of antimatter are exactly equal to each other. At sufficiently high temperature, the amount, or number density, of each species of particles and likewise of each type of antiparticle is of the same order as the number density of photons. That is the rule of thermal equilibrium. If energy can be transferred from one form to the other (and indeed it can, in our best current theories) then the total energy will be equally shared between the various possible forms it can take, be it massive particles or radiation. At lower temperature, the number density of very massive particles may differ from that of photons, and the differences are determined by the particle mass. Since we know that matching matter and antimatter particles have exactly the same mass, thermal equilibrium between them would say that their populations must be equal.

Now we do not know for sure that particles and antiparticles interact in such a way as to achieve thermal equilibrium with one another at high enough temperatures, but our best current theory says that they most likely do. So our question is now well defined: does an expanding Universe obeying the known laws of physics lead from our initial picture of matter and antimatter to the present one?

There are two important classes of processes that determine the history of matter and antimatter. First, whenever a matter particle meets its matching antimatter particle, they can both disappear by producing some form of radiation, for example two photons. We call this process annihilation. Second, there is a reverse process, whereby radiation, say a collision of two high-energy photons, produces a matter particle plus its matching antimatter particle.

There is an important difference between these two types of processes, annihilation and production of matter–antimatter. Whereas annihilation can take place whenever a particle and antiparticle collide, even slow-moving ones, production can happen only if radiation with sufficient total energy collides. Specifically, Einstein's mass-energy relation $E = mc^2$ plus the law of conservation of energy tell us that the radiation must have more energy than twice the mass-energy of a particle in order to be able to produce that particle together with its antiparticle. (Any extra energy becomes motion energy of the produced particle and antiparticle.)

In the hot early Universe, the photons were very energetic, so in a typical collision the total energy was well above two times the mass energy of any particle. Production of matter and antimatter was possible, and indeed as likely as annihilation of a particle and antiparticle to form radiation when they collide. Thermodynamic equilibrium then tells us that the number of matter and antimatter particles must be similar to the number of photons. Trying to keep a lot more energy in photons than in matter–antimatter pairs is like trying to maintain a mixture of two gases with a different temperature for each type. It cannot be done: collisions redistribute the energy as uniformly as possible; that is what we mean by thermodynamic equilibrium.

Our statement that the number densities of particles and of photons were of the same order of magnitude in the early Universe is therefore entirely independent of the initial state. But, as the Universe expanded, it cooled. Consequently, the population of radiation with sufficient energy to make massive particles and antiparticles decreased. Annihilation remained a significant process, but production slowed down. Eventually, when the Universe was cold enough, production stopped (or became so rare that we can forget about it). For protons and neutrons, this happened when the Universe was about a millionth of a second old. The temperature dropped below about $10^{12}$ kelvin, and most photon collisions were not energetic enough to produce proton–antiproton or neutron–antineutron pairs.

From then on, if pairwise production and annihilation are the only possible processes, there was a one-way trend: particles and antiparticles were steadily disappearing to make photons, and there was no process that could replenish them. So eventually the antiparticles would disappear, but along with them all the particles too. The Universe today would then be a very dull place, no stars, no planets, no people, just diffuse radiation and dark matter. Clearly that is not the case; after all we are here!

This is where the mystery of the missing antimatter begins. How did it happen that the antimatter, when disappearing from the Universe, did not take along with it all matter too, leaving behind only radiation? The real mystery is not the disappearance of the antimatter, but the fact that some small fraction of the original matter got left behind.

We must have left something out, some additional piece of physics must intervene in the history of the Universe, allowing antimatter to disappear, but leaving behind a small amount of matter of which present-day galaxies are made. So now we need to take a close look at what physicists understand about matter and antimatter, and their interactions, in order to address the question of how an imbalance between them could arise.

# 4 WHAT IS ANTIMATTER?

## What Is Matter?

> When I use a word it means just what I choose it to mean . . . the question is which is to be master.
>
> *Humpty Dumpty*, arguing with *Alice* (Lewis Carroll,
> *Through the Looking Glass*, chap. VI)

To talk about the physics of matter and antimatter, and even dark matter, we need first to define these words as physicists use them. We physicists have a tricky habit of redefining words as we learn more about the world. Like Humpty Dumpty, we assume that the ability to redefine the meaning of a word depends who is the master. (Many of us even forget we have redefined a word, expressing surprise when others fail to grasp the concepts described so precisely by our specialized and restricted physical or mathematical meaning of a word.) In everyday usage words typically have a wide range of meaning, but a physical quantity must be precisely defined to be useful, so physicists continually change and refine word meanings as they learn more.

The word "matter" is one whose definition has evolved, of necessity, as we have learned more about the underlying structure of the world. Eventually physicists found it convenient to redefine the word quite dramatically. Still, all the substances originally conceived to be matter are seen as matter under the new definition. But the old definition is simply inadequate to comprehend the newer discoveries, and in particular to accommodate the concepts of antimatter and dark matter.

Originally matter simply meant anything with mass. Even today you will find matter defined in many middle-school textbooks as that which has mass and which occupies space. For most of what we experience in everyday life this definition, properly interpreted, still suffices. It could be refined, as it was by the beginning of the twentieth century, or by high-school text books today, to "all matter is made of atoms" or perhaps "atoms are matter." While the second statement is true, the first is not; atoms are not the whole story. Let us explore further.

Everyone plays peek-a-boo with a baby. Why is the game such fun? Laughter often comes from happy surprises. But why is the baby surprised to see your face again? The baby is just beginning to recognize a fundamental law of physics, one you take for granted, the property of persistence of atoms. When Alice grows or shrinks after taking a bite from her magic mushroom you know this is a fantasy, not a story of actual experience. This is because atoms have a third property; in addition to having mass and taking up space (that is, having volume, even though tiny), atoms have the property of persistence. They do not just appear and disappear, except in the case of radioactive decay, where one atom can disappear although even then it is replaced by a collection of others of roughly equal total mass.

This recognition is so basic to our experience of the world that its formulation as a conservation law in science is lost in prehistory. Greek philosophers had some concept of it as they developed the ability to weigh and to measure the density of matter, skills important for merchants and traders as well as philosophers. Certainly the alchemists of the middle ages were well aware of it, though they tried to challenge its limits. They tried to transmute lead to gold, starting from a dense substance, because even they did not expect to get mass from nothing.

Philosophers may debate the question of what it means to exist, as they have done in many ways, in many cultures. But from a scientific point of view, matter is something that exists because its presence is measurable, and because the measurements can be repeated consistently. It has properties which persist.

High school chemistry teachers teach the persistence of matter as a law of conservation of mass. They lie, but only a little. This was the early observational form of the law; we will see that it is not precisely true. However, approximate laws that are very close to being true are important too. Often it takes some time before the deeper and more precise form of the law can be understood, and meanwhile the approximate one is very useful. Both in the history of ideas and in the education of a student one cannot deal with all the complexities at once.

Every science teacher lies over and over again in this sense. They state simplified versions of the laws of nature as if they were exact, to avoid the confusion that ensues when all the ifs and ands and buts are stated. In this book we do it too; though for the most part we will at least give you clues when we are doing it—if you look out for those weasel words "almost," "approximately" or "to a good approximation," and "in a certain limit" you will notice where we are fudging the truth, not telling the whole story.

Even conservation of mass took some work to establish; it is much more abstract than the persistence of solid objects. Methods of scientific observation had to be developed and refined. When one burns firewood it appears that the mass decreases; when a plant grows its mass increases. One must account for invisible forms of matter, keeping track of what flows in or out of the visible system, the gases given off by the fire, the water and carbon dioxide absorbed and oxygen emitted by the growing plant. This recognition was an important and necessary step in formulating the persistence of matter as a scientific conservation law.

Later recognition of further, and more abstract, conservation laws required a key understanding that was gained here. One must either study an isolated and contained system, or one must carefully track what flows into and out of the system, in order to be able to observe what is or is not a conserved quantity. Truly isolated systems are hard to produce, so usually the latter approach must be taken.

It sounds obvious, doesn't it? But notice that already "observation" does not mean simply the same thing as "seeing." Observation includes any repeatable form of measurement. The step from simple visual observation to inferences based on quantitative measurements of unseen things will lead us down the rabbit hole and into the wonderland of particle physics. Devising the right form of measurement or observation is an important part of the science.

It is not very long before the chemistry teacher gets to a much better form of the law of persistence of matter. Once atoms and their chemical reaction equations are introduced, the persistence of matter takes a more explicit form. Atoms are never created or destroyed in chemical processes. The same number of atoms of each element must be present after any reaction as before it. The conservation of mass turns out to be an approximate result that follows from this fact, plus the fact that the mass of any molecule is approximately, though not exactly, equal to the sum of the masses of the atoms that it contains.

Let us explore that "not exactly" for a moment. The mass of any molecule is actually a tiny fraction less than the sum of the masses of the separate atoms. Even though each isolated atom is electrically neutral, there are electric fields inside them. Molecules form when the energy stored in these fields can be lowered by combining the atoms in overlapping arrangements. The missing mass $m$ between the molecule and the sum of the atoms is given by $E = mc^2$ where $E$ is the binding energy of the molecule. That is, the energy that must be provided to take apart the molecule into its constituent atoms. This missing mass is typically about one part in a billion of the mass of the molecule. Of course when the molecule was formed and that mass disappeared the energy did not, it simply went into some other form, such as motion energy of the molecule.

This tiny fraction, however, is the key to why the molecule is stable; its total energy is not enough to allow it fall into pieces. Here we are using another conservation law which overrides and corrects the law of conservation of mass, which is the law of conservation of energy. Mass is just one of the many forms of energy according to Einstein's famous $E = mc^2$ relationship. (There is a common misconception, fed by the approximate law of mass conservation being stated as an absolute, that Einstein's relationship between mass and energy applies only in nuclear physics. On the contrary, it applies for any isolated system, no matter what forms of energy are contained within the system.)

In chemical interactions a collection of atoms starting out as one set of molecules are rearranged into a different set of molecules. The energy needed to make a reaction happen (for an endothermic process), or the energy produced by such a reaction (for an exothermic one), corresponds precisely to the difference in the sum of the masses of the molecules before and after the reaction.

Notice that the law of conservation of energy would make it impossible to have chemical reactions that produce heat if the law of conservation of mass were indeed a law of chemical interactions—you cannot get thermal energy, or any other form of energy for that matter, from nothing. Notice too our emphasis on conservation laws as the critical laws that help us predict what can happen and what does not happen. Such laws are important tools.

To be fair to the chemists, they never talk about the mass of a molecule. There is a mass that is constant; it is simply the sum of the masses of all the atoms before and after a chemical process. The fact that one can define the mass of a molecule and that it is slightly less than the sum of the atom masses does not change this conservation, which is actually conservation of atoms in chemical reactions. The difference between the mass of the molecule and the sum of the masses of the atoms it contains (times $c^2$) is called the binding energy of the molecule. It is tiny compared to the sum of atom masses. For chemistry it is inconvenient to mix the scale of atom masses into the problem, since all the chemical action is at the scale of binding energies. So, wisely, the chemists leave the masses out of the energetics of their studies and talk only about changes in binding energy.

Now we take another drink from Alice's magic potion, and become so tiny we can observe the structure within the atom. Fortunately, we don't really need to shrink ourselves; we simply need tools that allow us to make observations that are sensitive to this structure. That is what the huge accelerators and detectors of a high-energy physics laboratory do for us (figure 4.1).

As we enter the world within the atom, seeking to understand and explain the variety of atoms and their properties we find new refinements—atoms are made of protons, neutrons, and electrons. So we agree we must call all these things matter particles, since they are the constituents from which all our previously defined matter is made. Neutrons were first fully understood in 1932. Once they were, elements could be defined by the number of protons and their different isotopes understood as having different numbers of neutrons in the atomic nucleus. To make a neutral atom the nucleus must be surrounded by a cloud of electrons,

Fig. 4.1 That is what huge accelerators and detectors do for us.

with the number of electrons matching the number of protons in the nucleus. Chemical properties could also be understood, determined by the number and arrangement of electrons in an atom, particularly by the number in the outermost (or highest-energy) occupied states for a neutral atom of each element. Quantum mechanics was the tool that made all this more than a set of rules of thumb, codified in the periodic table. It explains the structure of that table and is the basis of modern physics and chemistry.

All the understanding of more modern physics of matter, with its proliferation of particle types, and eventually a substructure for protons and neutrons, came about as physicists tried to understand what are the forces between protons and neutrons that cause them to form nuclei, and to explain the radioactive decays of unstable nuclei. It turned out that the struggle to answer to these questions required a huge detour and the detour became the field of particle physics. But our modern understanding of particles and their interactions does much more than provide the answers to the questions about nuclei; it opens up whole new arenas of study. Cosmology, or history for the Universe, is one of them.

## Dirac Introduces Antimatter

No one worried about the mystery of the missing antimatter before 1928. Indeed, no one had even imagined such a thing as antimatter until a mathematical equation, written by a young mathematical physicist named Paul Dirac (1902–1984; Nobel Prize 1933), then a Fellow of St. Johns College at Cambridge University, predicted that it must exist.

Even Dirac himself took a while to recognize and accept what his equation was predicting. This equation was a crucial step in our modern understanding of physics. Stephen Hawking said, in a memorial address for Dirac, "If Dirac had patented his equation, like some people are now patenting human genes, he would have become one of the richest men in the world. Every television set or computer would have paid him royalties."

Paul Dirac was the son of a French father and an English mother. He grew up in England. He was a very quiet man, and very focused. It is unlikely it ever occurred to him that he might have patented his equation; he was not a particularly worldly person. After his work had been awarded a Nobel Prize a London paper described him as "as shy as a gazelle, and as modest as a Victorian maid"—not your usual image of a physicist, though we know a few like him. Indeed he considered turning down the Nobel Prize because he did not want the public attention it would bring, till his friends told him that turning it down would bring even more attention.

But that was later. When he found his equation he was trying to produce a formulation that combined quantum mechanics and Einstein's special theory of relativity. There was a known way to do this, but Dirac simply didn't like it. He wanted a particular mathematical feature that it did not have. His reasons for this were not very clear, but his quest led to an equation of the form he sought. His equation could describe the interaction of electrons with magnetic and electric fields; and it gave correct values for certain properties of the electron never before understood. (The earlier formalism works perfectly well for some other types of particles.)

The new feature Dirac achieved was that his equation would correctly treat half-integer spin. Spin is a weird quantum effect whereby particles have intrinsic angular momentum (a physicist's term for rotational motion) even though they do not really spin around themselves. In quantum theory, angular momentum, like the energies of radiation at a certain frequency, comes in discrete amounts. It can only appear in amounts that are *integer* (or whole number) multiples of Planck's constant $h$, the same quantity that gave the right black-body spectrum. This in itself is weird enough. But certain particles, the electron among them, were found to have spin 1/2 in units of $h$, something that you just cannot get by spinning anything around an axis. This turns out to have the consequence that you cannot put two electrons in the same state at the same time and place. This is known as the exclusion principle; it is a key to all of chemistry, which arises from the patterns of the filled states of electrons in atoms. But all this was only understood some time after Dirac's pioneering work.

In 1928 Dirac presented his new mathematical equation and showed that it could be used to calculate the quantum behavior of electrons, with or without surrounding electromagnetic fields. With this equation, now known as the Dirac equation, he made a dramatic change in the concept of matter, because, much to his initial surprise and even chagrin, it introduced the concept of antimatter.

One indication that Dirac had the right equation was that it correctly predicted a property called the magnetic moment of the electron. This quantity determines some aspects of how electrons respond to magnetic fields. It was well measured but not at all understood. Dirac's equation told him what it should be, and that matched the measured value. As he later said: "That was really an unexpected bonus for me, completely unexpected." For Paul Dirac that is a remarkably effusive statement.

Here is another story that suggests just how shy Paul Dirac was—in 1929 he and another young physicist named Werner Heisenberg (1901–1976; Nobel Prize 1932) (also one of the founding fathers of quantum theory), were traveling together on a ship. Heisenberg noticed that Dirac was not participating in the social life of the ship. Once when Heisenberg came to sit beside Dirac after dancing, he reported that Dirac asked him: "Heisenberg, why do you dance?" Heisenberg replied: "Well, when there are nice girls it is a pleasure to dance." "Heisenberg," protested Dirac, "how do you know beforehand that the girls are nice?"

But back to Dirac's equation, which turned out to be much more revolutionary in its predictions than its author had suspected. In addition to its success, the equation had some very strange properties, which took some time to be understood. As well as a negatively charged particle, the electron, the equation automatically required that there be positively charged particles too. This was unavoidable, and quite a surprise to Dirac. At first he tried to interpret these particles as protons, since these were the only positively charged particles then known. Indeed, protons and electrons were the only two types of matter particle known at that time, so this seemed the only option.

Dirac actually recognized that his equation insisted that the positive particles must have exactly the same mass as the electron, but he hoped that somehow that could be patched up. He speculated that the effect of electrical interactions between charged particles, interactions which he knew he had left out, would fix the problem. "I have not yet worked out mathematically the consequences of the interaction," he said, "one can hope, however, that a proper theory of this will enable one to calculate the ratio of the masses of protons and electrons."

Here he was totally wrong; even great physicists make wrong guesses. The identity between the masses of the two particle types cannot be removed by such effects, as he later realized. Dirac was, first and foremost, a mathematical physicist, and try as he might to avoid it, he gradually saw that there was no way to get rid of the mathematically imposed exact symmetry between the masses and other properties of the positive and negative particles in his equation.

Furthermore, by 1930 a catastrophic problem for this interpretation of Dirac's theory was recognized by Oppenheimer (1904–1967). Oppenheimer was then a young assistant professor at Berkeley, but already well versed in the quantum world. He later became the leader of the Los Alamos project to develop a nuclear bomb. Working independently, the Russian nuclear physicist Igor Tamm (1895–1971; Nobel Prize 1958) also pointed out the same problem. What Oppenheimer and Tamm saw was that Dirac's equation predicted a process in which two oppositely charged particles, one of each type described by the equation, could disappear if they came close enough to each other, leaving behind only photons. Physicists call this process "annihilation."

This would be a disaster. If the proton were the positive particle in Dirac's equation then this process would wreak havoc with the persistence of matter. If electrons and protons could annihilate one another, no atoms would be stable. Even a simple hydrogen atom could disappear into a puff of electromagnetic radiation. Since many types of atoms are stable, the proton cannot be the positive particle matched with the electron in Dirac's equation.

Given this situation, the outlook for Dirac's equation seemed dark. But eventually, in May 1931, Dirac had the courage to claim he knew what the equation was predicting (his way of saying it later was that he "made a small step forward"). He suggested that the negatively and positively charged particles in his equation indeed have equal masses. Thus the equation predicted a new kind of particle, which he eventually called an antielectron. Even then he hedged his bets, saying that this particle "*if there were one*, would be an entirely new kind of matter." The emphasis here is ours. His equation said a new type of particle must exist, but it seems he was not quite ready to bet on it.

## Experiments Confirm That Antimatter Exists

Today particle physicists postulate new particles at the drop of a hat, because, by now, we have seen that so often that is the best interpretation of evidence. Then, this seemed an entirely preposterous thing to do. So it took three years for Dirac to be convinced this was what his equation predicted. Only six months after that Carl Anderson (1905–1991; Nobel Prize 1936), studying events produced by cosmic rays, announced that he had seen evidence of this new phenomenon, namely, he had observed a particle with a mass much less than that of a proton but a positive electric charge.

In 1932 there were no high-energy accelerators. Anyone who wanted to do experiments studying high-energy particles had to do so by looking at the tracks of particles produced in the atmosphere by the impact of very high-energy particles arriving from somewhere out in space, known as cosmic rays. Anderson and Milliken (famous for his oil drop experiment which showed that charge comes only as whole-number multiples of the charge of an electron) had built a cloud chamber with a strong magnetic field in order to study the energies of cosmic ray particles by measuring the curvature of their tracks caused by the magnetic field.

In a cloud chamber, any charged particle leaves a trail of condensation droplets, formed as a consequence of the ionization caused by the passage of the charged particle through the gas in the carefully prepared chamber. Photographs are taken to record the patterns of these tracks. The first surprise was that, if all cosmic rays were coming from above the chamber, the curvatures indicated that there were positive as well as negatively charged particles.

The curvature of the track also measures its momentum, and this, together with the density of droplets, could be used to estimate the particle mass. That eliminated any possibility that this was a proton track. It had to be a particle with a mass similar to that of the electron. But that left open the possibility that they were electrons somehow coming from below the chamber—those tracks would curve in the same way as positive charges coming from above. Anderson had an idea how to tell in which direction the particles were going. He put a sheet of lead in the center of the chamber. A particle could lose energy by interacting with the lead as it passed through, but it could not gain energy. So the change in energy would tell the direction of travel. He soon found a picture that showed a particle that entered the chamber with high energy, then passed through a lead plate and emerged on the opposite side with lower energy. The curvature of the tracks showed that this was a positively charged particle feeling the force from the magnetic field. The loss in energy as it passed through the plate was shown by a change in curvature of the track. With the charge of the particle so determined, and the mass estimated, Dirac's antielectron was the only possibility. Anderson named it a "positron" and the name is still used.

Dirac's equation said even more. Not only do these new types of particles exist, but, given sufficient energy, a particle plus its matching antiparticle pair of any type could be produced, for example from photons and electric fields. Anderson soon saw events of this type too, where a pair of oppositely charged but equal-mass particles seemed to appear from nothing, near to the track of a high-energy charged particle moving through his chamber.

Physicists call this process pair production. Nowadays this mechanism is used at accelerator laboratories to produce beams containing millions of antiparticles, either positrons or antiprotons. Of course, for each antiparticle we also produce a particle, but we use magnetic effects to separate the two types and very high vacuum to ensure that most of the antiparticles we produce do not meet any matching particles to annihilate them before we can use them in our experiments.

It was soon realized that Dirac's equation should apply to any spin-1/2 particle. That meant it was making an amazing prediction. Not only the electron, but any half-integer spin particle, must have a mirror twin, a doppelgänger particle with identical mass but opposite charge. Protons and neutrons have spin 1/2, so they too must have equal mass and oppositely charged partners. (We will return later to the question of what it means to have opposite charge for a neutron.) The detection of the antiproton took a little longer, because it needed the right accelerator to be built. It was first achieved in 1955 by Owen Chamberlain (1920– ; Nobel Prize 1959), Emilio Segre (1905–1989; Nobel Prize 1959), Clyde Wiegand, and Thomas Ypsilantis in an accelerator known as the BeVatron in Berkeley. In another experiment in the same facility, in 1957, a team consisting of Bruce Cork, Glen Lambertson, Oreste Piccioni, and William Wenzel established the existence of the antineutron. (In one of the odder twists of Nobel history, in 1972 Piccioni sought damages for being wrongfully excluded from the 1959 Nobel Prize, claiming he had designed some of the experimental tools that made that work possible; his case was thrown out because of a statute of limitations. Nobel Prize rules allow a maximum of three recipients, and almost any experimental scientific effort today involves many more than three people in various aspects of the work.) More recently, antihydrogen atoms (made of an antiproton and a positron) have been produced and detected in experiments. The antiparticles of ordinary matter particles are all, by now, well-established entities.

The story of antimatter is a real departure from the old concept of persistence of matter. Massive particles appear from nowhere, given sufficient energy. They always appear in a paired arrangement, matching a particle with its Dirac-equation equal-mass partner. Such a new discovery needed new language—the new partner particles to the known matter particles were generically called antimatter. Slightly trickier usage evolved with regard to the word antiparticle; the new types of particles were the antiparticles of their equal-mass partners; but the already known particles are also called the antiparticles of their newly discovered antimatter doppelgängers.

In inventing the term antimatter for the new types of massive particle in Dirac's equation, physicists, probably without even thinking about it much, were drastically redefining the word matter. Previously matter was all that had mass; mass being one of the indicators for substance. Now we have two types of substance, two classes of particles with mass: matter particles, the proton, neutron, and electron; and antimatter particles, their Dirac mirrors, the antiproton, antineutron, and antielectron (the first discovered antimatter particle, also called the positron, it would be the only antimatter particle to be given a special name of its own).

The world of physics is a little like the land behind the looking glass. Physicists, like the Queen of Hearts playing croquet, feel justified in changing the rules of the game whenever they see a better set of rules. Of course, we are not quite as capricious about it as the queen. In fact, as the history of the Dirac equation showed, we physicists tend to be very stodgy and to hold onto our ideas as long as we can make them work, meanwhile speculating about what the next set of rules might be. But eventually nature forces our hand, by producing a result that is inconsistent with our theory. Even then, the old rules never quite vanish; they almost always stay around as an approximate set of laws, good for most everyday situations, but known to be incomplete when all evidence is included. The discovery of antimatter did not change the laws of chemistry. It simply added new options, new processes outside the bounds of chemistry. Atoms are still conserved in all chemical reactions, but should an atom meet an antiatom then watch out—they may both disappear in a flash!

However, a new conservation law took the place of the old one, since the particles and antiparticles only ever appeared or disappeared in pairs, always with a matched particle and antiparticle. Thus, even including the possibility of antimatter, there was a conservation law. The number of particles minus the number of antiparticles of any given type remained constant. Instead of conservation of number of protons, neutrons, or electrons, we now have conservation of the number of protons minus that of antiprotons, neutrons minus antineutrons, electrons minus antielectrons, and so on.

## Radioactive Decays of Nuclei

Meanwhile, in the same period, direct studies of nuclei broke down another key conservation law, the notion that the number of atoms of a given type cannot change. A radioactive atom is one that can spontaneously disappear, or decay. Early observations of radioactive decays of atomic nuclei identified three kinds of radiation occurring in such decays. Physicists named the radiated particles alpha, beta, and gamma. That is just Greek for *a*, *b*, and *c*; it simply says they had no idea what these radiations were. But today we know that the alpha particle is a helium-four ($^4$He) nucleus, a fragment left over after a larger nucleus fissions, and the gamma particle is a photon, carrying off energy from a transition of a proton within the nucleus from a higher-energy state to a lower-energy one. The most peculiar case is the beta decay. The beta particle is an electron coming from the decay of a neutron to a proton. The name beta decay for nuclear decays that produce electrons has persisted, but in more modern terminology this is a weak interaction. In either alpha or beta transitions the atomic nucleus present before the transition disappears and one or more different nuclei appear.

While alpha decay is just the break up of an unstable heavy nucleus to produce a helium nucleus among the fragments, beta decay is quite another thing. It is a revolutionary concept. A neutron turns into a proton and in the process produces an electron—the particle that exists before the process disappears and new particles that simply were not there before appear. This is unlike any chemical rearrangement of atoms or indeed anything else we know except at the quantum particle level. This was the first such quantum process observed, so of course it took a while before it was understood. As we develop the story of particle physics this type of process plays many important roles: we see it happen all the time in our experiments, so you will need to get used to the strange idea that one particle turns into a set of quite different particles when it decays.

But if neutrons can turn into protons in beta decays, and if pairs of protons and antiprotons or neutrons and antineutrons can be produced, what remains of the law of conservation of matter? All that is left is the conservation of a sum: number of protons plus neutrons minus antiprotons minus antineutrons. It is useful to introduce a new term here. Physicists call this number the *baryon number*. We classify both protons and neutrons as baryons, and antiprotons and antineutrons as antibaryons. The baryon number is the number of baryons minus the number of antibaryons. Said another way, each baryon carries a label +1 for baryon number and each antibaryon carries a label −1 for baryon number. Baryon number, like electric charge, is a quantity that is conserved in all processes that have ever been seen in any of our experiments.

Furthermore, the conservation of the number of electrons minus anti-electrons is also lost when an electron is produced in the beta decay process. (We know they are by observation, and also we know the conservation of electric charge requires some negative-charge particle to be made when a neutron becomes a proton.) So this law too must be replaced with a more general sum. The answer turns out to be that another new type of particle called a neutrino (or rather its antiparticle) is produced along with the electron, and the conservation law now involves the number of electrons plus neutrinos minus their antiparticles. Again we introduce a new terminology to go with this rule: we call both the electron and the neutrino *leptons*, and our rule is conservation of lepton number, the number of leptons minus the number of antileptons. The existence of neutrinos and the form of this rule took a little time to be recognized, because neutrinos interact so little that early beta decay experiments had no way to detect their presence. So we now turn our attention to Wolfgang Pauli, the person who figured out this piece of the puzzle.

# 5 ENTER NEUTRINOS

## Pauli: The Beta Decay Puzzle

Wolfgang Pauli (1900–1958; Nobel Prize 1945) was not a man to be indefinite about things. Born in Austria and educated in Germany, he spent most of his working years in Zurich, Switzerland. He was a man who held strong opinions, and had no time for fuzzy thinking. The sight of him in the front row of the audience for a physics talk, shaking his head and muttering "ganz falsch" (completely wrong), caused many a speaker some nervous moments. Pauli had played an important role in many aspects of the development of quantum mechanics. He was a very active participant in the network of physicists working to understand quantum ideas and radioactive decays. His name is perhaps most familiar in connection with the exclusion principle, the rule that says that you cannot put two identical spin-1/2 particles, such as electrons, in the same state at the same place and time. This rule is known to anyone who studied chemistry, as it governs the possible states of electrons in any atom. Pauli showed that this was a consequence of the mathematics of spin-1/2 theories.

Pauli was well aware that there was a problem with beta decay. In comparing nuclear beta decays to gamma decays, physicists found a big puzzle. In any given beta decay process they saw a continuous spectrum of energies for the electron, but nuclear transitions that gave gamma particles (or photons) always gave certain discrete energies, suggesting a discrete set of nuclear energies. How could these two observations be consistent?

One possibility was that the laws of conservation of energy and momentum had broken down. These laws had worked in all other cases. However, in starting down the road of quantum theory, the physicists of the time had to give up many ideas that they had held before, so some were willing, if reluctant, to explore the possibility that these laws too were an approximate set of rules that did not hold exactly at the level of nuclear processes.

An alternate explanation was first suggested, in letters and in private conversation, by Pauli around 1930. Pauli saw that the idea that nuclei have a discrete set of possible energy states and the fact that the electron emitted in beta decays had a continuous spectrum of energies could be made consistent with the laws of conservation of energy and momentum (and indeed of angular momentum too) if another particle was being produced at the same time as the electron. With only two final particles these conservation laws give only one possible electron energy for a given atomic transition, but with an additional particle there are many ways to share the available energy around among the three and still achieve conservation of momentum. The newly added particle had to have a very tiny mass, so as not to violate conservation of energy for events with the highest-energy electrons seen, and it had to have spin 1/2 so as to conserve angular momentum. It also had no electric charge, because that was already correctly accounted for by the proton plus electron.

To Pauli is seemed that introducing this new and effectively invisible particle was a better option than giving up three conservation laws. These conservation laws, as we will later describe, are deeply rooted in one of the basic assumptions of physics, the assumption that the laws of nature are invariant with respect to time and place. It seems that inventing a new particle, revolutionary as that idea may be, was less of a revolution to Pauli than abandoning such well-grounded conservation laws. Pauli, however, cannot have been all that excited about this idea. He sent a letter to the organizers of a meeting ("Dear Radioactive Gentlemen") which included this idea, but apologized that he could not come to discuss it personally as he had to go to a ball, where his presence was indispensable.

The idea that you could patch up a basic law by postulating an unseen particle still seemed quite preposterous to many. Such a particle could not be detected by any test one could make at that time, and it was not part of any ordinary objects. Furthermore, at that time most physicists still thought that nuclei were somehow made from protons and electrons; the idea of neutrons was still not broadly accepted. From that point of view it seemed quite natural for electrons to be emitted with a range of energies. The quantum theory, with its strict energy levels in the nucleus, was not yet part of the thinking of all physicists.

## Fermi: The Theory of Neutrinos Develops

Enrico Fermi (1901–1954; Nobel Prize 1938) took Pauli's suggestion seriously. Fermi was a leading physicist in Italy, until the rise of fascism in that country caused him to leave, immediately after receiving his Nobel Prize. Fermi soon became an intellectual leader of the U.S. nuclear physics community, and is famous for his experiment that was the first to observe controlled nuclear fission. The term *fermion*, which means any particle with spin 1/2, is based on his name, because of his contribution to understanding the statistical properties of collections of such particles. These properties are deeply related to Pauli's exclusion principle, which tells us that no two fermions can be placed in the same state. In contrast, particles with integer spin values are called bosons, after the Indian physicist Satyendra Nath Bose (1894–1974), who recognized certain important properties that arise from the statistical behavior of these particles, which unlike fermions can populate a single state many times over—and so do not have an exclusion principle. (This property is called Bose–Einstein statistics, because Bose had to enlist Einstein's help to get his work published.)

Fermi named Pauli's unseen particle a *neutrino*—little neutral particle, using the Italian diminutive—and built this idea into the working theory of beta decay by 1934. He recognized that a neutrino, a particle with spin 1/2 but no mass and no electric charge, must be produced along with a positron whenever a proton made a transition to a neutron inside a nucleus (a beta-plus decay, which occurs in certain proton-rich nuclei). In the more common neutron-to-proton transition of beta-minus decay, an electron is produced along with an antineutrino. Fermi developed a mathematical theory of such decays that could predict in detail the spectrum of electron energies from beta decay. This was true to form for Fermi: his theoretical ideas, while deep and novel, contained predictions that could be immediately tested in experiments. Whenever he could, he made these tests before he published his ideas.

We have already remarked that the observation of beta decay required generalization of the rule of conservation of particles minus antiparticles. Beta decay obviously removes separate conservation of proton number (protons minus antiprotons) and neutron number (neutrons minus antineutrons), but an overall conservation law for the sum of these two quantities, that is, for baryon number, is still viable. Likewise, in the Fermi theory of beta decay, we find conservation of (electrons minus antielectrons) plus (neutrinos minus antineutrinos), conservation of overall lepton number, but not the two species separately. In this process an electron is produced without its matching antiparticle, but instead with an antineutrino. Indeed, this is why we chose to call that particle the antineutrino and the one produced in beta-plus decay a neutrino; that naming preserves the pattern that matter and antimatter can only be produced in pairs. Conservation of electric charge is also maintained in beta decay; at the same time a neutron turns into a proton an electron appears; when a proton turns into a neutron, a positron appears.

Notice that beta decay introduces a radically new concept, along with the new particle, the neutrino. A neutron disappears and three new particles appear, particles that were not there before: a proton, an electron, and an antineutrino. When we talk about decays in particle physics we are almost always talking about processes such as this, processes where the things that arise when something disappears are not objects that were in any sense inside the initial particle. Indeed, we find many examples of particles that have several different and mutually exclusive ways of decaying—falling apart sometimes into one set of particles and sometimes into another. In quantum theory this can all be comprehended, but only as a set of probabilities. A quantum theory of particles can tell us the half-life for the particle to decay and the relative likelihoods of finding each possible final state in a collection of many such decays, but it can never tell us which decay will happen in a given instance, nor when the next decay will occur.

## Cowan and Reines: Neutrinos Detected

The introduction of the neutrino allows the conservation laws of energy, momentum, and angular momentum to be correct in beta decay processes as in all others. But does this particle really exist? It was not easy to figure out any way to detect neutrinos. Since they have only weak interactions (and gravitational ones) they are very elusive beasts. These particles were not detected until more than twenty years after Fermi's theory was developed and widely and successfully used to describe nuclear beta processes.

To have any chance of detecting neutrinos one needs a very intense source of them, such as a nuclear reactor, and also a very large detector. Neutrinos are then detected when they are absorbed by a neutron, producing a recoiling proton and an electron. (Likewise, antineutrinos can be absorbed by a proton to produce a neutron and a positron.) The tracks of the recoiling charged particles suddenly appear in the middle of the detector. Something must have entered the detector to produce this effect. If the experiment is set up with enough shielding material and veto power (detection capability) to exclude any other particle coming in, then the entering something can only be the (anti)neutrinos produced in the reactor. This experiment was first done by Clyde Cowan (1919–1974) and Fred Reines (1918–1998; Nobel Prize 1995), who worked on the problem for several years, starting in 1951 at Los Alamos. They eventually succeeded in demonstrating the effect quite convincingly at Savannah River in 1956. Immediately they sent a cable to Pauli to tell him his idea had been confirmed. Why it took so many years before the Nobel Prize was awarded for this pioneering work is one of the mysteries of Nobel history; by the time it occurred Cowan was no longer alive to share the prize and indeed Reines was barely well enough to know his work had finally been recognized. Perhaps the delay was because by the time the experiment succeeded most physicists would have been very surprised if there was no neutrino. It seems one gets less recognition for confirming an expected but difficult to detect particle than for finding one that is not so precisely expected.

This experiment also demonstrates the long chain of inference that can be necessary to interpret particle physics experiments. When physicists say they have detected a new particle, the evidence is often quite indirect. Physicists search for, and find, evidence of certain processes that cannot be explained without the new particle and are easily explained if such a particle exists. A case begins to build. The physicists then make further predictions for what should occur if the particle exists. They say they have seen the particle if enough cases fit their predictions, and none goes counter to them. After a few such particle discoveries, physicists quickly lost the reticence that Dirac and even perhaps Pauli had felt in postulating a new particle—in the 1950s and early 1960s, the data from accelerators showed evidence that required more and more of them. We will talk of some of the others in our next chapter; all of them, in one way or another, are players in our story.

A number of experiments have recently demonstrated that neutrinos have tiny, tiny masses. Until a few years ago physicists thought they may have no mass at all. This discovery opens an intriguing possibility; it may well be that these elusive particles hold the key to the imbalance between matter and antimatter in the universe. We need much more background before we can tell that story; we will do so later in this book.

# MESONS

## Yukawa and the Pi-Meson

Before we can describe a very important ingredient that was missing in our attempt to follow the cosmological history of matter and antimatter, we need to get to know a few more of the particle participants in this story, beyond the particles that we have encountered so far (protons, neutrons, electrons, and neutrinos and their antiparticles).

It did not take long after the idea of antimatter and its experimental verification for the particle story to become even more complicated. In 1935, a remarkable Japanese physicist, Hideki Yukawa (1907–1981; Nobel Prize 1949), predicted an entirely new type of particle which he called *mesons*. He argued that such particles must exist as mediators of the strong nuclear force (the force that binds protons and neutrons together in the atom's nucleus).

The nature of this force was then the central puzzle of nuclear physics. Once it was understood that nuclei are composites made from protons and neutrons it is clear that there must be some force that binds them together. It is neither electromagnetism, which pushes the protons apart and barely affects the neutrons, nor gravity, which is much too weak to provide the glue to bind such compact objects as nuclei. The studies intended to answer this question were to lead to lots of new discoveries, including many types of particles, before they finally led to a satisfactory answer. Indeed, they led to the entire discipline known as particle physics,

Yukawa must have been a remarkable man. Given the traditional culture of Japan at the time, there were relatively few scientists there who were educated in Western science and who were following the developments of nuclear physics, and even fewer who tried to do original work. Yukawa was a leader in a small group of up-to-date physicists at the University of Kyoto. Without modern means of communication his connection to his colleagues around the world by mail must have been frustratingly slow, yet still he was a key player in the development of ideas. His influence was strong enough that Heisenberg and Dirac even went to Japan to visit him (the journey by sea where Heisenberg danced and Dirac asked him why).

The pattern of stability of nuclei and other data had given clear indications that the nuclear force, while very strong at short distances, falls off in strength very rapidly as nucleons are separated, much more rapidly than do electrical forces. Yukawa first tried to explain nuclear forces via electron exchange, but he soon realized that this could not give such a short-range force. That, and a number of other problems of the electron-exchange theory, made it unworkable.

Yukawa recognized that he had to postulate a new particle to mediate the strong force and that it should have no spin. He needed a set of three of them, all with about the same mass, one neutral and two with electric charges +1 and −1. Exchange of these particles between protons and/or neutrons could then provide a theory of the strong nuclear force that correctly reproduced major observed properties of that force. These particles must thus also themselves have strong interactions with protons and neutrons (and with one another).

Yukawa saw that the range of the force could be used to predict the mass of these exchanged particles. They had to be a lot more massive than an electron but a lot less massive than a proton. (A force mediated by a massive particle gets weaker with distance in a way that depends on the particle's mass.) He also noted that the only place such a particle might be directly produced was in a high-energy nucleon-nucleon collision. He suggested that high-energy cosmic rays colliding with atoms in the upper atmosphere should produce many of these mesons, which are now known as pi-mesons, or pions.

To discover a particle that is not present in everyday matter, you need two things, a way to produce them (natural or artificial), and a way to detect them. Until the advent of high-energy particle accelerators (effectively until the 1950s) one had to look for natural sources. Cosmic rays are high-energy protons (and possibly other types of particles) that constantly bombard the earth from space. These highly energetic particles were discovered by their ionizing effect on the upper atmosphere (which affects the propagation of radio waves). They were named cosmic rays by meteorologists who flew instruments on meteorological balloons to measure the altitude dependence of ionization in the atmosphere.

These balloon-carried experiments began around 1911 and continued into the 1930s. The incoming particles were at first thought to be high-energy photons, which is why they were named cosmic rays, but eventually it was recognized that their latitude dependence could only be explained if they were charged particles, since their trajectories were affected by the Earth's magnetic field. This conclusion was confirmed by the fact that, in laboratory experiments, the cosmic ray particles that reach the surface of the earth were found to penetrate with high probability through a couple of inches of gold. Photons cannot do that.

Nowadays we distinguish between the *primary* cosmic ray particles that arrive at the top of the atmosphere, which are chiefly very high-energy protons, and the *secondary* particles produced in the collisions of the primary particles with atoms of the atmosphere, and even *tertiary* particles, which are the decay products of the secondary particles.

In the 1920s and 1930s, physicists began to use particle detectors, first developed to study emissions from radioactive sources in the laboratory, to study cosmic rays. Early particle detectors came in three basic types: emulsions, cloud chambers, and Geiger-type counting devices. Emulsion detectors were photographic-type materials that could be exposed to radiation and then developed, much like a photograph, except here the picture showed tracks left by a charged particle traveling through the emulsion. Cloud chambers were devices prepared so that energetic charged particles left a track, like the contrail of a passing jet plane, made from condensation of droplets on the ions created by collisions of the charged particle with the gas in the chamber. These tracks were then recorded photographically. (We have already mentioned that Anderson discovered the positron using such a device.) Counters were similar to the Geiger counter still in use today to detect sources of radiation. They register the passage of a charged particle that ionizes the gas in them by producing a small pulse of electric current for each such event and, in their original version, transforming it into an audible click. (Nowadays an electronic counter is more common.)

The search for Yukawa's meson first turned up something completely unexpected. Tracks seen in the emulsions fitted the expected behavior for a charged particle with about the mass Yukawa had suggested. But physicists soon realized these tracks were from a particle with no strong interactions. They traveled long distances through matter without being deflected by interactions, whereas a strongly interacting particle such as Yukawa had predicted would soon be deflected and quickly absorbed in matter.

These new particles (which have electrical charge, either +1 or −1) were eventually named *muons*. Muons are particles that are very much like electrons (or for the +1 charge, positrons). They have all the same types and strengths of interaction as do electrons and positrons. The only distinction between muons and electrons comes from the fact that the muon is much more massive, and is thus unstable, decaying by a weak interaction to give an electron and two neutrinos (or rather a neutrino and an antineutrino). Such a particle was a complete surprise. No one had predicted anything like that.

Eventually, by 1947, cosmic ray emulsion experiments also found the tracks that were made by Yukawa's particle, nowadays called a pion (or a pi-meson). Large numbers of pions are produced by cosmic ray primary particles colliding with the atoms of the upper atmosphere. Since the pions decay, or are absorbed in the emulsion via strong interactions, their tracks are quite short. Such tracks were discovered in emulsions flown high in the atmosphere. Charged pions decay due to weak interactions. In the decay, the pion completely disappears and a muon and a neutrino (or an antineutrino) appear. Thus the charged particles, or cosmic rays, that reach the surface of the earth, are almost all muons.

Yukawa's particles, the pi-mesons (or pions), although they have mass, have integer rather than half-integer spin. They were only the first of many types of mesons to be found, the least massive. All types of mesons can play a role in nuclear processes. They are all unstable particles, produced and observed in high-energy collision experiments, but rapidly disappearing to produce less massive particles. (Eventually only electrons, protons, and neutrinos are stable, and neutrons too, but only when bound in stable atoms.) Mesons do not behave according to the Dirac equation, because they have integer spin. (The earlier equation, the one that Dirac did not like, works perfectly well for them.) So are they matter, or antimatter, or something else again? As we will explain later, the best answer is neither matter nor antimatter but a mixture of both. This was again something quite new; and it took some time to sort it out.

Accelerators were invented in the late 1940s as tools to study the interactions of protons and neutrons and learn more about the forces between them that Yukawa had sought to explain with his meson. Much to everyone's surprise, they opened a Pandora's box full of new particle types—not just one type of meson but many different ones, different masses, different detailed lifetimes, and so on. Not only were new mesons unearthed but also new proton-like particles, a veritable zoo. We do not need here to explore the whole zoo, but we do need to take a look at a few of the beasts that turned out to have properties that are particularly important for our story.

## Strange Mesons, Strange Quantum Concepts

One particular type of meson, called *K-mesons* (or *kaons*), plays a special role. Some peculiar quantum properties of the neutral K-mesons are the basis of our first evidence that matter and antimatter do not obey completely matched sets of physical laws. These particles were discovered and their properties elucidated in bubble chamber experiments at accelerators. (A bubble chamber is a detector in which charged particles leave a trail of bubbles that can be photographically recorded.) There are in fact four particles now called K-mesons, two with electric charge ±1 and two with electric charge zero. These particles were dubbed *strange* particles quite early on because their properties did not match those of other mesons; they were much longer lived than pions, but more massive, and that seemed strange. Normally one would expect the heavier particle to decay faster, as more energy is released in its decay.

Particle properties are first discovered empirically; then we make up rules, conservation laws, that can "explain" what we observe and what does not happen. We then ensure that our theory enforces these rules. Just as we can write a theory so that it gives us the observed property of conservation of electric charge, so we can invent other kinds of charges (labels carried by particles) and specify when they are conserved and when they can change.

The unexpectedly long life of the K-mesons could be encoded as a new type of charge, one that is conserved in strong and electromagnetic interactions but can disappear in a rare type of weak interaction. (Remember that weak interaction is what we call beta decay, but it also gives pion decay and, in this rarer form, K-meson decay.) Strange particles were observed to be always produced (via strong interactions) in pairs, one with a +1 strangeness quantum number (charge) and one with a −1 strangeness quantum number. These particles are both unstable, and they decay individually. This means the decay does not respect this strangeness quantum number, it changes it. This is a new idea, a conservation law that applies for some types of interaction but not for others. It turns out to be a very useful concept in sorting out the patterns of particle decays. The strangeness-changing process is a weak interaction, like beta decay, but it is even rarer. The idea of introducing this quantity called strangeness, and further development and successful predictions based on this idea, earned the 1969 Nobel Prize for Murray Gell-Mann (1929–). We will meet him again later in our story.

# 7
# THROUGH THE LOOKING GLASS

## What Physicists Mean by the Term *Symmetry*

We have described many of the important particles that play a role in the story of matter and antimatter. Now we need to review some other aspects of particle theories that are important in this story. Remember that our history of the Universe contained some critical assumptions. We assumed that the early Universe had equal numbers of particles and antiparticles. We took into account that matching particle–antiparticle pairs can annihilate into pure radiation at any temperature but can be produced from pure radiation only when the temperature is high enough. We found that this theory said that the present Universe should have neither galaxies nor antigalaxies. Since we observe a huge number of galaxies, this story must be wrong.

A crucial point in this chain of logic is that we considered only two types of processes that affect the numbers of particles and of antiparticles, namely, pair production and pair annihilation. These were the only two such processes allowed by the Dirac equation. We have already found some variants of this picture, weak decay processes that change a particle of one type into one of another type, neutron into proton, and produce a particle and a not-quite-matching antiparticle, an electron and an anti-neutrino. Stranger yet, we find that mesons can completely disappear in weak decays, leaving only a muon and a neutrino. Surely the Dirac equation does not give the full picture. Perhaps there are other relevant processes. Perhaps the conservation law, conservation of number of matter particles minus number of antimatter particles, is simply not true. If so we need to extend our theories to allow processes that break this pattern, and test these theories by looking for them.

The Dirac equation has another feature that is crucial in this story. The laws of nature that are encoded in the equation are exactly the same for matter and for antimatter. Thus we say that the equation has a *symmetry*; we call this symmetry *CP*. This symmetry is another obstacle in obtaining a consistent history of matter and antimatter in the Universe. If the symmetry is correct, then no imbalance between matter and antimatter could develop, even if there were no conservation law that forbids it. Perhaps there are other relevant processes, ones that do not respect the CP symmetry. But physics is first and foremost an experimental science. If we want to know how to get beyond the restrictions of Dirac equation, we should test the notion of matter–antimatter symmetry through experiments.

What are symmetries? How do we test them? Does the CP symmetry, that is, the matter–antimatter symmetry, implied by the Dirac equation, survive experimental tests? These are the questions we need to examine before we can proceed further with our story of the Universe.

## A Gedanken Experiment

In Alice's experience, the world on the other side of the looking glass is a rather strange place, following laws quite different from those of the everyday world. But is this the reality of physics, or just the imagination of a clever mathematician author? This is the question of whether mirror reflection is a symmetry of nature. At first glance you might guess that it is, because nothing you see in a mirror looks physically impossible.

Here is a proposal for an experiment to test this question more carefully. Imagine a racetrack with two cars in the competition. But these are not just any two race cars. They are mirror images of each other. Whatever is on the left side of one is on the right side of the other. Even the drivers are mirror images of each other (figure 7.1). Most important, the engines are mirror images of each other. The cars are at rest, side by side. Then, at precisely the same instant of time, each of the drivers starts to accelerate

Fig. 7.1 These are not just any two race cars. They are mirror images of each other.

her car. Of course, one does it with the right foot pressing a gas pedal on the right and the other presses with her left foot a gas pedal on the left. The race is along a straight section of the racetrack. Will there be a winner? Or will the cars cross the finish line at precisely the same time?

Imagine that the cars indeed cross the finish line at the same time. Imagine further that we conduct more and more experiments of this type: we prepare initial conditions that are just a mirror reflection of other initial conditions and test whether each final situation is just the mirror reflection of the other. If the results were the same for every such mirror-image pair of experiments we could devise, we would say that there is mirror symmetry in nature. Such equality in the results would say that the equations governing the physics must have an *invariance* in them; if $x$ denotes the distance to the (imagined) mirror, changing $x$ to $-x$ would not change the equations.

This then makes predictions for all future experiments. For example, left-right inversion invariance in the equations describing the physics of the cars would lead to the prediction that the two left-right mirror-image race cars would finish at the same time. These equations would predict that, if there were another Universe which is just the mirror reflection of our Universe, the laws of nature there would look just the same as in ours.

When physicists use the word symmetry they mean just such an invariance in the equations that describe the physical system. The system has mirror symmetry if the equations that describe it are invariant under a reversal of the coordinate describing distance from the mirror plane.

The strong, electromagnetic, and gravitational interactions indeed have symmetry, called *parity*, denoted by P, that corresponds to mirror reflection, plus a rotation by 180 degrees about an axis perpendicular to the mirror. This reverses not just left and right, but also up and down and forward and backward. In other words, parity is equivalent to the impact of three mutually perpendicular mirrors; it reflects in all three dimensions of space. Since all known forces have invariance under rotations, the new feature of this symmetry is indeed just mirror reflection.

One aspect of this subject might be confusing. Many of the phenomena that we observe in nature are not symmetric under mirror reflection. In our brain, the functions of the left and right hemispheres are different. Our hearts do not lie in the middle of our body. DNA molecules have a left-handed spiral. One may wonder how we can even think that nature is left-right symmetric when there are so many phenomena that are not. A variant of an old story illustrates the answer. Once upon a time there was a donkey that was extremely thirsty. It was given two buckets of water. The buckets were put at precisely equal distances from the donkey, one to the left and the other to the right. The situation was manifestly symmetric under mirror reflection, which made things difficult for the poor donkey. To slake its thirst it had to choose from which bucket to drink (or, at least, from which one to drink first). Given the exact symmetry, the donkey just could not decide. There was no advantage whatsoever in choosing one bucket over the other. So the donkey ended up dying from thirst.

Not a very probable story, right? In spite of the exact symmetry, any donkey would have made a choice. The existence of a symmetry would not manifest itself in the death of the donkey but rather in the fact that the probabilities for it choosing the left- or the right-side bucket are exactly equal (assuming an unbiased or symmetric donkey).

In the physicists' language, the fact that a choice is made in an equal probability situation is described as *spontaneous symmetry breaking*. All the phenomena that we described above are of this type. Like any real donkey, nature sometimes makes a choice between left and right. But this is a manifestation of spontaneous breaking of mirror reflection symmetry; it does not contradict the fact that the basic forces that govern molecules are electromagnetic and, therefore, parity invariant. Nature could have chosen to have our brains with left and right hemispheres interchanged. Our hearts could dwell in the right side of our body. DNA molecules could be right handed. None of the functions of these objects would have changed. Like the donkey, nature had to make a choice, and it did. But the choice could have gone either way. The self-replicating property of DNA ensured that once this choice had happened, quite by accident, it would be reinforced by appearing forever in the replicated molecules. Thus we observe spontaneous breaking of parity in our everyday life, but all experimental evidence shows that the everyday interactions are parity symmetric. The two forces we experience in everyday life—electromagnetic and gravitational—have exact mirror symmetry. Based on this experience, physicists guessed that all interactions have this property.

But the question of whether parity is a symmetry of *all* the basic interactions has an experimental answer, and the answer is *no*. Of course, this answer was not obtained by conducting our proposed car race experiment. That is impossible to carry out. In particular, we cannot find two drivers that are exact mirror images of each other and can start their cars at the same instant. It is still useful to imagine such an experiment. Physicists often discuss such *gedanken experiments*, experiments carried out in thought rather than in reality. (*Gedanken* is German for thought.) While we cannot get answers from gedanken experiments, they help us to make the questions very clear, to expose paradoxes or shortcomings of our theories, and eventually to think about realistic experiments that would answer the questions at hand. The process that was first used to observe parity violation was not a car race but rather a particular example of nuclear beta decay.

## The Actual Experiment

The actual experiment that first looked for parity-violating effects was carried out in 1956, by C. S. Wu (1912–1997) and collaborators. While

Wu did not share the Nobel Prize awarded to the theorists Lee and Yang for the suggestion that led to her work, she was widely recognized as one of the leading experimental physicists of her time. In 1975 she was the first woman to be elected President of the American Physical Society, the U.S. professional association of physicists (a position that Helen held, as the fourth female President, in 2004). Wu's experiment studied the process in which a cobalt atom, $^{60}$Co, decays into a nickel atom, $^{60}$Ni, plus an electron and an (anti)neutrino. The trick here is that $^{60}$Ni has one unit less spin than $^{60}$Co.

Now when the decay occurs with the electron going off parallel or antiparallel to the original spin direction, then the antineutrino must go in the opposite direction—that is the only way you can have conservation of momentum and angular momentum in such a decay. Conservation of angular momentum also has a further consequence in this case. It says that the outgoing electron and neutrino must both have their spin lined up in the same direction, the direction of the spin of the cobalt before it decayed.

Thus, in order to conserve angular momentum, that is, to account for the spin lost from the nucleus in this decay, the direction of the spin of the antineutrino must be parallel to its travel (right handed) for any antineutrino that goes in the direction along which the cobalt spin has been aligned. Similarly, the spin would have to be antiparallel to the direction of travel (left handed) for an antineutrino that comes out in the direction opposite to the cobalt spin direction. Of course, this is true for the electrons as well. (Why do we call the neutrinos left or right handed? The terms are used because our fingers can curl only one way about our thumb, so if you imagine a thumb pointing along the direction of travel of the particle, you will understand which way a left-handed or a right-handed particle is spinning.)

Now the argument we need next depends on the fact that the direction of a rotation, and thus of spin, does not change under parity. So first you need to understand this point. To see this, consider a clock on the south wall of a room, in the usual orientation, with the 12 at the top. Its hands move in the clockwise direction about an axis through the center of the clock face. (That is, of course, what defines the convention for *clockwise*.) Now imagine that you have a weird clock which runs backward, with the reverse direction of motion of the hands, switching 3 and 9,

Fig. 7.2  If you picture it carefully, you can see . . .

etc., that is, reversing left and right on the clock face. Imagine you place the reversed clock on the opposite wall, putting the 12 down, with the center of the clock exactly opposite the first clock. You have now inverted all three directions. But, if you picture it carefully, you can see that the hands of the upside-down reversed clock are moving in the same direction about the line between their centers as those of the normal face-up clock—the direction of rotation of the clock hands is unchanged after this inversion (figure 7.2). (Again, a gedanken experiment; you do not need to go to the trouble to do it, but it is useful to imagine it.)

Spin is a form of angular momentum, so the direction of a spin is not reversed by parity. However, the directions of travel of the neutrino and the electron are. Thus the initial state of our cobalt atom is transformed by the action of a parity transformation into exactly the same state—a cobalt atom with its spin in the same direction as before. But the final state with the electron traveling off parallel to the spin direction is transformed by the parity inversion into a final state with the electron traveling in the opposite direction. Of course, the neutrino direction gets reversed too; it is always back to back with the electron. So, if the theory has parity symmetry, one must find equal rates for the two cases, electrons parallel to the cobalt spin or electrons antiparallel to the cobalt spin. Any difference between these two rates would show that the parity symmetry does not hold in weak decays.

The experimenters, led by C. S. Wu (respectfully called Madame Wu, even by her usually informal American physicist colleagues), set up a situation where the spins of all cobalt atoms were aligned in a known direction. This is easy to do by applying a strong magnetic field and then turning it off. They then counted the rate of electrons in the direction parallel to the spin of the $^{60}$Co nuclei and the rate for the opposite direction. The rates were indeed different. The emission of electrons favored the direction opposite to the nuclear spin. Indeed, it was found that the electron is never emitted in the direction parallel to the spin. The violation of the relationship required for a parity-symmetric interaction was as big as it could be.

To produce an electron parallel to the cobalt spin would require that the antineutrino went directly opposite to the spin and, as we argued earlier, it would then have to have its own spin aligned with the direction of the cobalt spin, and this would be a left-handed antineutrino. None were produced. Nuclear beta decay is mediated by the weak interactions. The results for $^{60}$Co decays, and for all subsequently observed weak processes, are beautifully explained if weak interactions act only on left-handed neutrinos, but they have no power to produce or affect the right-handed ones. In addition they act only on right-handed antineutrinos, and do not affect or produce left-handed ones.

So nature is not invariant under parity. More precisely, of the four known fundamental forces, three have exactly the same behavior when all directions are reversed. These are the forces of gravity and of electromagnetism, and the strong interaction. But weak interactions are different. They act differently on left-handedly spinning neutrinos and on right-handedly spinning neutrinos.

In the imaginary race that we described above, the outcome would depend on what force drives the engine. If it is based on gravitational, electromagnetic, or strong interactions, the two cars would arrive at precisely the same time to the finish line. There can be no winner. But if we could construct the engine so that the driving force is the weak one, then the two cars may accelerate at very different paces, and would cross the finish line at very different times. Indeed, if we choose a neutrino driver, one of them may never advance toward the finish line at all. (This should not worry you when you actually buy a car: all existing engines use processes that are driven by the electromagnetic interactions and therefore left-oriented cars are neither slower nor faster than right-oriented ones.)

Physicists had assumed for a long time that parity is a symmetry of nature. This prejudice was perhaps a result of the fact that parity is a symmetry of space itself, as well as a symmetry of all the common everyday processes we observe. Perhaps the boldest move toward the realization that weak interactions violate parity was to dare to pose the *question* of whether parity is a symmetry of nature. In 1956, T. D. Lee (1926– ; Nobel Prize 1957) and C. N. Yang (1922– ; Nobel Prize 1957) were the first to ask the question and to demonstrate that there was, up to that time, no reported evidence for or against parity conservation in weak interactions. The experimental test, with its surprising answer, came shortly after. (Rumour has it that some experimenters had seen parity-violating effects before this time but had been reluctant to publish such odd results when all the theories that they knew did not accommodate them—such timidity is bad science, but it certainly happens.)

Indeed, often the most challenging step is to pose a good question. Sometimes it requires you to step outside the borders of thinking and patterns of your culture, or of a scientific subculture. We are all very much captives of collective concepts. It takes special minds to break free. Originality often consists of asking a new question; once such a question is asked, new answers are likely to be found. Lee and Yang, two Chinese physicists working at Columbia University in New York, came together at that time to discuss particle physics and their discussion produced the right question. That earned them a Nobel Prize in 1957 and with it, near legendary status in the Chinese community.

Helen says: "I have enjoyed some truly excellent Chinese meals when T. D. Lee came to visit us at SLAC. When any Chinese restaurant was told that he would be our guest, they really went all out to do their very best! When I accepted an invitation to China in 1983, there was some complication in the process of arranging the details of the trip. I happened to mention this when talking to Frank (C. N.) Yang about something completely different; within days the problems had disappeared. I later realized, to my embarrassment, that my request to bring my children, then aged 9 and 12, with me to China and to travel with them and my husband to some tourist sites, as well as giving my official lectures in Beijing, had been more complex than my hosts could easily deal with. Their hesitation had been a polite way to refuse my request. I didn't get the subtle message. Once Professor Yang spoke on my behalf, I and my whole family were treated royally. That trip was a memorable one for all of us."

# 8 THROUGH THE LOOKING ANTIGLASS

## Another Gedanken Experiment

Weak interactions violate the symmetry called parity. But what does this have to do with the symmetry between matter and antimatter? Let us consider that case next.

Imagine another day at the races, with even more bizarre race cars: one of them is made of matter (and driven by a driver) and the other is made of antimatter (and, of course, needs an antidriver). We also make them mirror images of each other. We do so because an antimatter particle, in the sense of the Dirac equation, not only carries opposite charge to that of the matter particle, but would also be right handed if the matter particle is left handed and vice versa. To be concrete, let us call a car with a driver on the left side *left*, and one with the driver on the right side *right*. Let us further call the driving machine made of particles (protons, neutrons, and electrons) a *car*, and the one made of antiparticles (antiprotons, antineutrons, and positrons) an *anticar*. Thus we conduct a race between a *left-car* and a *right-anticar*. Now the driver of the second car is a right-handed antineutrino. Of course our cars must be rocket cars, as we will need to hold this race in space so as not to bias the outcome with a matter track. Will there be a winner?

The transformation of all charges to opposite charges is called *charge conjugation*, represented by C. (The experiments that discovered that parity is violated in weak interactions also showed that charge conjugation is violated by these interactions. On the other hand, like P, C is a symmetry of the strong, electromagnetic, and gravitational interactions.) The combined transformation of both C and P is called CP. Two objects that transform to each other when directions are all reversed *and* all charges replaced by their opposites are called *CP-conjugates*. Our left-car and right-anticar are CP-conjugates, as are their imagined drivers. So in our next race we want to test whether CP is a good symmetry of nature. This is a symmetry between matter and antimatter that could survive even though P and C separately are not symmetries of all processes. Physicists were prejudiced by their past—and by the mathematics of the theories that they then understood—and assumed it was so. Indeed the experiments that originally observed P violation saw a corresponding C violation so that it appeared from these results that CP was still a good symmetry.

To understand the actual experiments that were performed to test the question of CP symmetry, it is convenient to think about a variation on our race. Let us tie our right-anticar to the left-car. If we put them side by side, facing the same way, they will move forward together. But we can also tie them head to tail, facing in opposite directions with, say, the left-car facing forward and the right-anticar facing backward and upside down. If CP is a good symmetry, the two (anti)engines will exert precisely the same force but in opposite directions, and the combined machine will move neither forward nor backward. If the combined machine moved in either direction, it would demonstrate CP violation. It would show that the laws of physics for particles and those for their parity-mirrored antiparticles are not the same.

The question that we are asking is whether the combined transformation of left to right, forward to backward, up to down, and matter to antimatter is a symmetry of the fundamental forces in nature. For the gravitational, electromagnetic, and strong forces, the answer is clear. Since each of P and C is a good symmetry for these interactions, then the combined operation, CP, must be a good symmetry also. But for the weak interactions the question is more subtle. Could it be that parity invariance and charge conjugation invariance are violated in a compensating way such that the combined operation is an invariance of the weak interactions also? Could it be that a left-car is faster than a right-car, and a left-car is faster than a left-anticar, but the differences are precisely the same, so that when we race a right-car against a left-anticar, they will arrive at the finish line at the same instant?

While you may find this possibility contrived, this is precisely what seemed to have happened in experiments that searched for parity and charge conjugation violations. The violations of P and of C by the weak interactions seemed to be related to each other. Weak interactions acted on left-handed particles and not on right-handed particles. Weak interactions acted on left-handed particles and not on left-handed antiparticles. But in all early experiments weak interactions seemed to act with precisely the same strength on left-handed particles and on right-handed antiparticles. Physicists were quite happy with this result; it suited their prejudice that CP is a good symmetry of nature. This prejudice was also supported by all the examples of field theories (that is the name of the mathematical theories that describe elementary particles and the interactions between them) that they then thought were likely descriptions of nature.


## Cronin and Fitch: Matter and Antimatter Do Not Follow the Same Laws

In 1964, an experiment to test a prediction made from CP symmetry was performed by James Christenson, James Cronin, Val Fitch, and René Turlay. Cronin (1931– ) and Fitch (1923– ) received the 1980 Nobel Prize in physics for their leadership of this work. James Cronin was born in Chicago, and is presently a professor emeritus at the University of Chicago. He is still an active experimentalist, involved in the Pierre Auger observatory, a giant detector array near the town of Malargüe in Argentina's Mendoza province. The experiment aims to solve yet another mystery, which is the nature and origin of rare but extremely energetic cosmic rays. Val Fitch was born on a cattle ranch in Cherry County, Nebraska, a highly improbable place (in Val's own words) to begin life that would lead to a Nobel Prize in physics. He is a professor at Princeton University. Fitch and Cronin met at Brookhaven, and later joined forces in Princeton University to study neutral K-meson decays. They fully expected not to see the process that they searched for, because CP symmetry would forbid it. Their very surprising result, that the process does occur, told them that the symmetry was broken. To explain what they saw we need to tell you a little more about the neutral strange mesons and their rather peculiar quantum properties.

Quantum mechanics brings not only new language, and new labels, it brings some entirely new concepts into physics. In classical physics, objects, such as particles, were things that had a definite mass and a definite (unique) substructure and constituents; a cherry pie always contains cherries, it is not sometimes full of raspberries and sometimes of cherries. If you take it apart you always get cherries. We have already seen that at the quantum level things decay, that is, fall apart, into objects that they did not originally contain, and that a particle can have more than one set of possible decay products. We now meet another odd quantum feature: a composite particle can have two or more mutually exclusive sets of constituents.

This makes the usage of the word *particle* another one of those where physicists take a common concept and have to refine it to describe in words what the mathematics of quantum theory is telling them. Is a particle something with definite substructure, or something with a definite mass? Classically, of course, it is both, but in the quantum world we must choose.

If we keep the name particle for objects that have definite mass and half-life, then these objects may have a composition that can only be described in terms of probabilities. In some sense this is the best choice, because an object of definite mass travels through space in a definite way, and hence these objects have an ongoing property that matches our concept of particle.

Every so often it is convenient to describe things in terms of the other set of states, the ones with definite composition; since we do not have another good word, we often call those particles too. Physicists freely switch back and forth between the two languages for the particles, as it suits them. They invent names that differentiate which description they are using; this can be confusing to the nonexpert, as the multiple names seem to suggest more particle types than there actually are. In this book we will generally use the word particle only for objects of definite mass and half-life, and use the term states to signify either particles or something that can be thought of as a definite admixture of two particles.

These two languages are both useful to describe the neutral K-mesons.

In either description there are two distinct states. The production of neutral K-mesons (which occurs in processes that involve strong interactions) is best described in terms of the states with definite strangeness. Remember that, like electric charge, strangeness is a label that reverses sign as you go from a particle to its antiparticle partner. Most particles carry a definite amount of this attribute, zero for protons, neutrons, and pions, +1 for charged kaons with a positive charge, and −1 for negatively charged kaons. Strong interaction or electromagnetic processes always make a matching particle and antiparticle and thus always conserve strangeness, just as they do electric charge. So far so good; but when we get to the neutral K-mesons we find a new twist. There are two states of electrically neutral K-mesons produced in strong interactions, these are called K-zero ($K^0$), with strangeness +1, and anti-K-zero ($\overline{K}^0$), with strangeness −1. Each of these states is the CP conjugate of the other.

However, when we want to discuss the decay of neutral K-mesons, we do better to talk in the language of particles, the states of definite mass and half-life. The definite mass states of the neutral K-mesons are called K-long ($K_L$) and K-short ($K_S$). These are particles that have a well-defined half-life as well as a definite mass. Remarkably, although they are similar in mass, the one we call K-long has about a five hundred times longer half-life for its decays than the K-short. (Half-life is the time in which half the particles in any given sample will have decayed.) This is the basis of their names, long for the longer-lived particle, and short for the short-lived one. K-meson decays can occur only via weak interactions. Weak decays do not conserve strangeness, although weak interactions that change strangeness proceed more slowly than those that involve only ordinary matter particles (those with strangeness zero). All that was understood and still the two neutral K-meson states with two very different half-lives were a puzzle.

Quantum mechanics tells us that the states of definite mass, $K_L$ and $K_S$, must be two different combinations of the $K^0$ and $\overline{K}^0$ states. Note, however, that the number of different objects remains the same, two definite-mass particles, or two states with definite strangeness. We can use one description or the other, but not both at once; there are only two things here. The crucial point is that if CP were a good symmetry of all interactions then each of the definite-mass and lifetime states, K-long and K-short, would have exactly equal probability of being K-zero or anti-K-zero. They differ in the relative sign in the coefficient of these two things in the quantum amplitude for the state, and this sign affects predictions for how the two particles can decay. (Think of this as our car and anticar tied together; the two having the same sign is like the cars facing the same way, their effects add, but if the two have opposite signs, it is like the cars facing opposite ways, their effects can cancel one another.) If CP is not a good symmetry, the states of definite mass can be any mixture of $K^0$ and $\overline{K}^0$; equal probabilities are not required.

CP symmetry also implies that the likelihoods for a K-zero or for an anti-K-zero to decay into two pions are equal. When combined into the states with definite mass, these contributions can add up, as they do in the K-short combination, or cancel each other, as they do in the K-long state. If CP were a good symmetry, this cancellation would be exact. The K-long meson could never decay to two pions. The K-long decay to two pions resembles the fate of our combined left-car and right-anticar machine that would not move unless CP is violated.

Indeed, the observed pattern, which had originally been mysterious, because it seemed odd to have two otherwise identical particles with such different half-lives, appeared to be explained by CP symmetry. The K-short is short-lived compared to the K-long because it can decay to two pions, while the K-long should not. All other possible decays for neutral K-mesons proceed much more slowly than the two-pion decay (a fact that is well understood). So the assumed CP symmetry predicts the two very different half-lives for the two definite-mass states.

But, if CP is violated, just a little, then the cancellation may be close but not exact. Two different half-lives would still be observed, but every so often a K-long would be seen to decay into two pions. That was the result of the experiment, and it was a shock. The experimenters found what they were looking for, but not expecting to see. (Experiments must test any prediction that something does not happen as carefully as they test those that something will happen.) Their result shows that CP is violated by some part of the weak interactions.

The violation was tiny. The rate for K-long to decay into two pions was found to be about a thousand times smaller than that for K-short. Such decays are very rare, but they definitely do occur. CP is not an exact symmetry of nature. There must be a small but significant difference in the laws of physics for matter and those for its mirror antimatter.

Helen says: "In 1964 I was a first-year graduate student at Stanford, studying particle physics theory, so this is the first discovery of this saga that I remember from my own experience. At the time I did not understand enough about field theories to know when and how CP symmetry could be built into or removed from the theory, but I do remember the amazement at this result that was expressed by faculty and in seminar talks. I knew something dramatically unexpected had been observed."

The discovery of CP violation had a profound impact not only on particle physics—new theories that would encode laws of nature that are different for matter and antimatter had to be formulated—but also on cosmology and, in particular, on the cosmological story of matter and antimatter. Some years earlier, in a letter to Heisenberg, Pauli had suggested that there should be a link between the observed matter–antimatter asymmetry of the Universe and an asymmetry in the laws of physics, but his remark seems to have been forgotten—probably because other problems were more pressing. But experiment forced attention to the problem of CP violation. Once it had been understood, physicists were ready to take on the challenge first suggested by Pauli so many years earlier.

# 9 *THE SURVIVAL OF MATTER*

## Pauli's Other Letter: Initial Conditions on the Universe

Wolfgang Pauli's letter to the "Radioactive Gentlemen" suggested the neutrino. Now another letter written by Wolfgang Pauli has a place in our story. He was, as far as we know, the first to discuss the fact that the observation of the positron raised a deep new question. Why is the world populated with electrons but not positrons? This cosmological issue is the central mystery of our book—why are we surrounded by matter but no antimatter? Pauli raised it in a remarkable June 1933 letter to Werner Heisenberg. He said: " . . . I do not believe in . . . (Dirac's) theory, since I would like to have the asymmetry between positive and negative electricity in the laws of nature (it does not satisfy me to shift the empirically established asymmetry to one of the initial state)."

This letter was written well after the experimental confirmation of the positron, which Pauli had certainly heard about, so Pauli is not denying the existence of antimatter. What he is saying is that he thinks there must be some difference in the laws of physics for matter and those for antimatter, to explain the observed difference in their populations in the Universe—which is what he means when he says "the empirically established asymmetry." In the Dirac equation, aside from the reversal of electric charge, the laws of physics for matter and antimatter are exactly the same. Modern-day physicists call this a matter–antimatter (or CP) symmetry in the equations; Pauli talked about positive (positron) and negative (electron) electricity. What bothered him was not that the Universe is asymmetric between matter and antimatter, but that the asymmetry exists when the laws of physics have a symmetry between the two species. It is enough to make him discount the Dirac equation even after its predictions have been confirmed by experiment.

Pauli also points out a possible answer to this puzzle, though he says it does not satisfy him. If the amount of matter minus the amount of antimatter never changes, then, and only then, the imbalance between them in the Universe can be explained by simply postulating that the Universe somehow began with such an imbalance, and, since it cannot change, it persists to this day. This is what Pauli means by "shift the asymmetry to one of the initial state." As far as we know, this is a first statement of the view, held today by most particle physicists and cosmologists alike, that it is unsatisfactory to appeal to initial conditions to explain the dominance of matter over antimatter in the Universe. Pauli also points out that, to avoid having to appeal to initial conditions, one must somehow remove the symmetry between the laws of physics for matter and antimatter seen in the Dirac equation.

However, the Dirac equation and its descendants, the modern theories of particle physics, had so many successes that, for about thirty years, physicists put Pauli's dislike of the symmetry between matter and antimatter to one side, and all their theories had such a symmetry. Only when experiments proved this symmetry was broken did they return to explore theories of the type Pauli suggests must be found, with the asymmetry in the laws of physics, and thereby to open up the door to new answers, other than an initial condition, to solve our mystery. Indeed, they needed to know much more before they could write sensible theories with no matter–antimatter symmetry. So the path they took, ignoring the cosmological problem of matter–antimatter asymmetry, was not so strange. They had more compelling problems in explaining direct laboratory observations. Pauli too put his reservations aside and pursued theories with matter–antimatter symmetry as actively as anyone else.

Let us dwell a little more on this notion of an initial condition. When we discussed the history of the Universe with matter and antimatter before, we assumed an initial condition that matter and antimatter were in thermal equilibrium with one another and hence had equal populations.

It could be that that assumption was wrong. So we need to look at that more closely.

First, we notice that if there is any process that can change the numbers of particles minus antiparticles, baryon number and lepton number, then our particular choice of initial condition is not an assumption, it is the condition that will quickly be achieved, no matter what the initial situation was. We know this by the rules of thermal equilibrium. Because particles and their antiparticles always have exactly equal masses, if there is any process that can change their relative numbers, any process that can change the total baryon number, then they will reach thermal equilibrium with exactly equal populations, and thus zero baryon number, as we assumed. In the hot early Universe this would have happened very fast. But if no baryon-number-changing processes exist then, whatever the baryon number one has at the Big Bang, that will persist. Likewise, if lepton number is conserved, then that too must be fixed by an initial condition on the Universe. Even if just the difference between these two numbers is conserved, that will allow us to pick a starting value that gives the observed Universe.

We cannot completely exclude this answer to our puzzle. However, there are three arguments that cause us, and most particle physicists and cosmologists, to have deep suspicions that it is not the right answer.

The first reason is that given by Pauli—it does not satisfy us. The main reason for this dissatisfaction is that the initial value of the difference between the number of particles and that of antiparticles that we must choose to get the observed Universe is very peculiar. It is a tiny fraction (approximately one in a hundred million) of the totals of either—such a fine-tuned condition does not seem natural. If we allow lots of finely tuned initial conditions we cannot prove that the Universe did not begin yesterday. You cannot investigate cosmology if you think that fine-tuned initial conditions may play a role in bringing the Universe to its present form. So we would prefer to avoid having even one fine-tuned initial condition.

If, however, there are exact conservation laws that say that the baryon number cannot change, then we have no choice. Like it or not, an initial condition would be the only possible answer. Our second set of objections thus come from the fact that, when we extend our best current theory in very natural ways, suggested by the similarity of the mathematical patterns of the different interaction types, we generally get theories where there is no such conservation law. We will explain this statement in more detail when we discuss grand unified theories. Indeed, even the current theory says that, in the hot early Universe, there are processes that violate the separate conservation laws of baryon number ($B$) and lepton number ($L$). The difference between them ($B - L$) is conserved, according to this theory, but only in a somewhat accidental fashion, unlikely to be maintained by any extension of the theory. This is also explained later, when we discuss symmetries. So we suspect there is no such conservation law, and then no initial condition will persist and we must solve the mystery in another way.

So far we have not seen any direct evidence for processes that can change either $B$ or $L$. Under conditions found in our laboratories we find that these quantities are conserved. Protons do not decay, though certainly we can think of possible decay processes that satisfy all other conservation laws. Indeed, searches for proton decay, which violates baryon number conservation, put a lower limit on the half-life of an isolated proton at greater than $10^{33}$ years, much longer than the age of the Universe. Clearly we need something very different to occur in the hot dense early Universe to get around that limit. If we are to extend our theory to break the conservation laws of baryon and lepton number, we must find a way to do it that allows these processes to occur freely in the hot early Universe but makes them exceedingly rare in the current conditions. This is not as hard as it might seem. For example, these processes could be mediated by extremely heavy particles.

The third reason for rejecting the possibility that the mystery is solved by fine-tuned initial conditions is that modern cosmology assumes that the Universe has undergone a period of extremely rapid expansion. This chapter in cosmology, known as *inflation,* has now very strong support from observations. The inflationary expansion dilutes all initial densities to incredibly tiny populations. Thus, as the Universe emerges from this period of extremely rapid expansion, it is practically empty of both baryons and antibaryons. It is subsequently repopulated with particles and antiparticles at high temperature, in a stage called *reheating* that occurs as the inflation stops and a slower rate of expansion takes over. But to have different populations generated from the reheating, the matter–antimatter asymmetry must appear in the laws of nature. In other words, inflation makes the actual initial condition irrelevant and the effective initial condition, the condition that emerges from reheating, is not a matter of choice, but of physical dynamics. At the end of inflation, the number of baryons and antibaryons is zero, so, in particular, the difference between them—baryon number—vanishes. Any imbalance must be generated by some physical difference in the laws of physics for matter and antimatter, whether it arises in the reheating stage or at some later time.

## Sakharov: The Conditions Needed to Develop an Imbalance

Now we know that in reality the laws of nature are slightly different for matter and antimatter. Could such a difference play a role in cosmology? Could it solve the mystery of the missing antimatter without imposing a finely adjusted initial condition?

The first person to address this puzzle seriously was the Russian physicist Andrei Sakharov (1921–1989; Nobel Peace Prize 1975) in 1967. This is about the same time as he began to write another famous piece of work, on a very different subject, entitled "Progress, peaceful coexistence and intellectual freedom." Sakharov was indeed a remarkable man. Together with Igor Tamm, his teacher, he was one of the developers of the Russian nuclear weapons. He later became an active voice for nuclear disarmament and for human rights in his own country. He was awarded the 1975 Nobel Peace Prize for this work, but was so unpopular with his own government that, together with his wife Yelena Bonner, he was exiled to the small town of Gorky from 1980 to 1986. There he was unable to visit his scientific colleagues, or go to a library to read the work of others, yet he continued to be a leading researcher.

Helen says: "During this time I was fortunate to host a small gathering at my home so that Sakharov's wife, Yelena Bonner, who had come to the United States for medical treatment, could meet with a group of physicists and talk about life in Gorky. Andrei too was not well at that time. I remember Bonner's fatalistic statement that their only future was that 'one day we will just quietly die'—and the implication was that this would not be due to their health. The next week an article written by

Bonner appeared in *Newsweek*, about her desire that people in the Soviet Union could live their lives as freely as she saw those in the United States did. Its title was 'I want a house.' However, it seemed to me that this house in her dreams might be mine, where Bonner had said, as she stood on the deck looking at the hills, 'It's so beautiful here, if Andrei were here he'd be well in a week.' None of us then imagined how soon the Soviet Union would crumble."

During the time of internal exile, a few of Sakharov's colleagues made the effort to travel to Gorky periodically, to bring him news and scientific papers, and to discuss scientific ideas with him. His release from Gorky came with the beginning of the period of Glasnost, and indeed he became a political figure in the new order in Russia after the collapse of the Soviet Union. Unfortunately he did not live long enough to achieve the nuclear disarmament he had so strongly argued for, though he did indeed help start his country along the path of nuclear detente with the United States, and of reduction of nuclear arsenals, with ongoing removal and destruction of these weapons.

Yossi says: "When Andrei Sakharov was at last allowed to visit the United States and came to SLAC, a series of short presentations on 'What is new in particle physics' was arranged for him there. I was then a young postdoctoral fellow in SLAC, and was asked to explain the importance of a new phenomenon that had just been measured whereby the two neutral B-meson states had been observed to oscillate one into the other. I felt so much respect for Sakharov, and was so surprised to find myself explaining physics to him, that I literally lost my words. This was by far the poorest presentation that I gave in my life."

But let us return to his groundbreaking 1967 paper. Sakharov enumerated three conditions that must be met for the imbalance of matter and antimatter to be produced, starting from equality, and to persist.

First, and most obviously, there must be an active process that can change the baryon number, that is the amount of matter minus the amount of antimatter. The baryon number is zero if matter and antimatter are in balance, but is certainly not zero today, as we see matter but practically no antimatter. To explain when the imbalance arose we need to find a time when processes that can change baryon number occur readily, but soon after they must become unlikely (because of new conditions).

Otherwise, whatever imbalance is produced will again be removed as the trend to thermal equilibrium enforces equality between matter and antimatter.

Second, the imbalance can arise only when the Universe is going through a transition stage that takes it temporarily out of thermal equilibrium, because otherwise, whatever is the process causing baryon number to increase, by the rules of thermodynamics the reverse process, which decreases baryon number, will occur at the same rate, and so no imbalance will develop. Suppose, for example, that among the baryon-number-changing processes there is a transition from an initial state $A$ to a final state $B$. In thermal equilibrium, the populations of $A$ and $B$ will be balanced so that the inverse process, $B \to A$, occurs at an equal rate. Thus, even if the particles in $A$ carry a different total baryon number from those in $B$, the balance of $A \to B$ and $B \to A$ processes prevents any change in baryon number.

We know of two ways to get around this, to avoid a thermal equilibrium situation in the Universe, and we will discuss an example of each. One is to produce some kind of massive but very weakly interacting particle, with the imbalance between particle and antiparticle types produced when these massive particles decay. If these massive particles interact rarely enough that they are not exchanging energy and momentum with any other particles around them, and if they decay slowly enough that by the time most of them decay the typical energies of radiation are not high enough to allow the inverse process of production, then they are not forced to maintain thermal equilibrium with the rest of the particle species. We say they drop out of thermal equilibrium when these conditions are met. The second way is to have the entire Universe goes through a phase transition. (The transition of water to ice is a phase transition.) During such a transition, the system has a region at any surface between the two phases that is not in thermal equilibrium. We will explain what we mean by a phase transition in the Universe by an example later. It is a remarkable idea, but it seems that indeed this did occur—for example at a stage of spontaneous symmetry breaking.

Sakharov's third condition is equally important: both C and CP symmetries must be broken. If the laws of physics were entirely symmetric between matter and antimatter, or even between matter and mirror antimatter, then any interaction which increases baryon number would be matched by a complementary reaction which decreases it (simply replacing all particles in the process by their (mirror) antiparticles).

In either case, if these symmetries were exact, no imbalance could ever develop. For any process that decreases baryon number—for example, baryon decay—there would be a process that increases it—in this example, the decay of an antibaryon (or mirror antibaryon)—which would occur at exactly the same rate. In our generic example above, for the transition we wrote as $A \to B$, exact CP symmetry would imply that the CP-conjugate process $\overline{A} \to \overline{B}$ (with $\overline{A}$ the set of mirror antiparticles of $A$ and $\overline{B}$ the set of mirror antiparticles of $B$) occurs at exactly the same rate. The final result would be disappearance of equal numbers of $A$ and $\overline{A}$ and production of equal numbers of $B$ and $\overline{B}$. Since $A$ and $\overline{A}$ carry opposite baryon number to each other, and so do $B$ and $\overline{B}$, net baryon number does not change. In particular, if the starting point is zero baryon number, that is, matter–antimatter equality, then the final baryon number is still zero. Clearly this argument works just as well if we apply C instead of CP, so we cannot have either symmetry if we want the baryon number to change at some time in history of the Universe. Indeed, we now know that nature has neither of these symmetries.

To understand the different roles of the three conditions, let us imagine a town in which there are an equal number of red and blue cars. We ask whether after some time the number of red cars could be different from that of blue cars. The absence of baryon-number-violating interactions is analogous to a situation where there are no roads connecting to this town. Without such roads, the numbers do not change. Even if the roads exist, thermal equilibrium would mean that the number of red cars arriving is equal to the number of red cars leaving town, and the number of blue cars arriving is equal to the number of blue cars leaving town. C or CP symmetry implies that the number of red cars arriving (or leaving) town is equal to the number of blue cars arriving (or leaving) town. Obviously, if a difference between the numbers of red and blue cars is to develop in the course of time, we had better have roads (baryon-number-changing processes), the total number of cars of at least one color arriving must be different from the number leaving (no thermal equilibrium), and the number of red cars arriving minus those leaving should be different from that of blue cars arriving minus those leaving (C and CP violation). We need all three effects.

Notice that, once we introduce any process that can change baryon number, it not only allows us to find an explanation for the occurrence of the imbalance between matter and antimatter, it requires us to do so. If any such process exists then, no matter what initial condition we choose to impose, the early Universe quickly achieves equal amounts of matter and antimatter, thermal balance between particles and antiparticles. That is why we chose equality of matter and antimatter as the starting point of our history for the Universe.

## Cosmology with Sakharov's Conditions Met: Baryogenesis

Suppose now that the three conditions are satisfied: there is some stage or period in the history of the Universe where some (anti)particles are not in thermal equilibrium, and these have interactions that violate baryon number, C, and CP. Let us run the clock forward from an initial state with equal amounts of matter and antimatter and see how the final picture changes in comparison to our earlier failed picture. Here we are interested in a qualitative picture only. We will come back to specific models for the critical stage later.

The important implication for cosmology from laws of nature that are different for matter and antimatter is that it is possible to arrive at a situation where the number of baryons is not equal to (say, is larger than) the number of antibaryons. If CP-conserving interactions are more probable than CP-violating ones then we would expect that the imbalance between the number of baryons and of antibaryons would be very small. What we mean by a small imbalance is that the difference between the number of baryons and that of antibaryons (the baryon number) is much smaller than the number of baryons or of antibaryons separately.

Now imagine also that, at this point in time, the changing conditions (such as the decreasing temperature) in the expanding Universe cause the baryon-number-changing processes to become extremely rare, so rare that a typical particle can cross the observable Universe before such a process occurs. Then after that time the baryon number really does not change any more. The small difference between the numbers of baryons and of antibaryons is "frozen in," guaranteed to be constant for the rest of the history of the Universe. For the sake of concreteness, let us assume that for every 1,000,000,000 antibaryons we now have 1,000,000,001 baryons.

The rest of the story here is not so very different from the one described earlier. For a while, baryons and antibaryons annihilate and are produced at similar rates, keeping their number close to the number of photons. But when the temperature drops below $10^{12}$ kelvin, most photons are not energetic enough to keep producing baryons and antibaryons. Then the number of both baryons and antibaryons becomes steadily smaller and smaller. When a baryon and a matching antibaryon meet, they can annihilate each other.

Finally, all antibaryons disappear through these annihilations. But not all baryons disappeared. For every 1,000,000,001 baryons, 1,000,000,000 have annihilated together with the matching antibaryons, but one is left behind. The small surplus of baryons created at almost the beginning of the Universe can find no antibaryons to annihilate with. They must be there forever. Can this be the matter we see today?

This final picture is a Universe devoid of antimatter. There is, however, a small amount of matter. This situation is a result of the small imbalance between the amounts of matter and antimatter that developed through CP-violating interactions in the early Universe. Whichever finds itself on the losing side in the matter–antimatter race disappears at the end of the story. The winning side has the reward of remaining, though in small quantities, forever. The smallness of the initial imbalance translates directly into the very small ratio between the number of baryons and the number of photons in the present Universe.

The facts that the laws of nature are different for matter and antimatter and that there are processes that do not conserve baryon number allow us to imagine a consistent history that starts from a simple initial state, with equal amounts of matter and antimatter, and ends up in a Universe with no antimatter and little matter. This chapter in cosmology where the imbalance appears is called *baryogenesis*. The violation of CP, demonstrated in experiment only in 1964, plays a crucial role in baryogenesis.

We emphasize that, unlike a tiny imbalance in initial conditions, there is no fine-tuning when the imbalance is dynamically generated, that is, induced by CP-violating interactions. We know that CP violation is a small effect in particle interactions. But there is still a challenge: does our theory give the correct imbalance or a different one? The test of a theory is whether it can produce the right answer, the ratio of 1,000,000,001 to 1,000,000,000 which is (in round numbers) what we need.

Cosmology or, more specifically, baryogenesis provides a qualitative explanation of the two striking features of antimatter and matter in the Universe, namely, the absence of the former and the small amounts of the latter. We are thus quite convinced that some mechanism such as baryogenesis, making use of baryon-number-violating and CP-violating interactions, is the source of the baryon asymmetry (the unequal amount of matter and antimatter) in the Universe. There is no need for contrived histories that rely on initial conditions.

Indeed, since even the Standard Model theory has baryon-number-changing processes, which are active at high temperature, it is unlikely that we can rely on any initial condition to solve the problem. We have to resort to conservation of $B - L$, baryon number minus lepton number. We do not know whether that is exact, but experience seems to show that every conservation law of this type that we postulate turns out to be broken when we learn more details, so we expect that this too is not an exact result. As we said before, once no conservation law protects it, any initial condition does not persist, and equality of matter and antimatter is quickly achieved in the very hot early Universe. The question that remains is whether our theories can account for the baryon number of the Universe quantitatively. Do we get the right ratio of baryons to primordial photons? An answer to this question can only be given in the framework of a specific theory. We will explore it shortly for the Standard Model but first we must develop a bit more of the picture of exactly what the asymmetry between matter and antimatter looks like in that theory. What are the features of the theory that allow it to appear when earlier theories had no such possibility? How do those features play out in the early Universe to allow an imbalance between baryons and antibaryons to arise?

# 10

*ENTER QUARKS*

## Quarks

Particle physics took a great leap forward in the 1960s with the idea that many so-called elementary particles—protons, neutrons, mesons, indeed, all particles that are subject to the strong nuclear force, collectively known as *hadrons*—might not be so elementary after all. By that time over a hundred types of hadrons had been observed in accelerator experiments; physicists had found many additional baryons beyond the proton and neutron, and many mesons in addition to the pion and kaon that we have so far introduced. The notion that all these different types of particles were distinct elementary building blocks of matter did not seem a very appealing picture. It was almost as bad as assuming that the atoms of each element were indivisible and structureless. The many possible elements, the different atomic nuclei, made much more sense once they were seen as composites made of protons and neutrons. Could one similarly explain the many known elementary particle types in terms of a few constituents?

By the early 1960s it was also known that, unlike electrons, protons and neutrons had a measurable size, with an internal distribution of charge. (Experiments that demonstrated this in some detail were done at Stanford with a forerunner of the SLAC accelerator, earning the 1961 Nobel Prize for their leader Robert Hofstadter (1915–1990).) But if protons and neutrons themselves, along with all the other hadrons, had substructure, what was it?

Fig. 10.1 You have developed a catalog of objects that have been found, and also one of objects that you looked for but could never find.

A simple game can help to illustrate how the important idea of substructure can arise without ever observing any of the constituent particles. We call this game a backward jigsaw puzzle, because here you are given the outcome, the pictures, and asked to discover the pieces, while in a jigsaw puzzle you are given the pieces and asked to construct the picture. Unlike a typical jigsaw puzzle, these pieces can be put together in a number of ways. So, as the physicist detectives in an imaginary two-dimensional world, you have developed a catalog of objects that have been found, and also one of objects that you looked for but could never find (figure 10.1).

Now your job is to figure out the constituents from which everything in this imaginary world is made. Equally important, you must discern the rules by which they interact with one another to form the objects you observe. So first, before you read further, think a while about the first question and jot down your answer.

We have tried this game with many audiences and they all quickly see that all the objects can be formed by some combination of squares and triangles with a common side length.

Now you are ready to tackle the second question. Here the puzzle of the things not found plays an important role. You can see that all these things could have been made from the same basic set of shapes, so that is why, once you guessed the right set of shapes, you looked for all these things. The fact they were not found is an important piece of information in deciphering the rules of interaction—what does not happen is as much part of what physicists must explain as what does happen. You are looking for a simple rule that explains both what you see and what you do not. Or, said another way, you are looking for patterns in the data that you can summarize with a simple rule.

The answer is not hard to find: every triangle must have one and only one neighbor and every square must have two and only two. All the allowed shapes follow this rule; all the unseen ones break it in some way. We can turn this pattern into a rule of interaction by stating it this way: Every triangle has one bond and every square has two, and all bonds must be saturated by linking each bond with one and only one other bond.

So physicists played this game—or rather the real-world version of it. Over many years of experiments the catalog of things seen and not seen and the patterns of their interactions and decays were tabulated. Finally, with enough information collected, the likely substructure could be guessed and then gradually, over ten more years, the guess developed into a theory, including all the types of interactions other than gravity, and the theory was tested in many many ways.

Within the categories of baryons and mesons, physicists noted that there were subgroups of particles with the same spin and similar mass. In addition to protons and neutrons, there were other known baryons by this time: a set of spin-1/2 particles, just like protons and neutrons, but with strangeness quantum number −1, and even −2, and a set of spin-3/2 particles. Mesons too came in sets; some had spin-0, some spin-1. Gell-Mann and, independently, the Israeli physicist Yuval Ne'eman, found certain patterns of charge and of mass differences and of the quantum number called strangeness within these groupings, and related these patterns to the mathematics called group theory. These patterns were then used to predict a new baryon type, called the omega-minus (with strangeness −3). It was soon observed. It took only one event that could be definitively interpreted as the production and subsequent decay of such a particle to give dramatic confirmation to the idea that these patterns had some deeper meaning.

This was the key that led to the idea that all hadrons, both the baryons and the mesons, were composite objects, constructed from a new type of fermionic, that is spin-1/2, particle. In 1964 Murray Gell-Mann (1929– ; Nobel Prize 1969), building on his earlier work that introduced the patterns of particles with different strangeness, recognized that all the known hadron patterns could be explained if there were three different versions or *flavors* of these new particles. About the same time, George Zweig made a very similar suggestion. Zweig called the new particles aces and Gell-Mann called them *quarks*, a nonsense word he took from James Joyce. The name quark is the one that is used today. Two of the quarks, now known as up and down, are found in protons and neutrons, while the third type, the strange quark, carries the strangeness quantum number that accounts for the properties of K-mesons and other recognized strange particles.

The two classes of hadrons and their different patterns of charge and mass had a simple explanation in terms of two allowed types of quark structure. Baryons—protons, neutrons, and any other spin-1/2 or spin-3/2 particles subject to the strong nuclear force—have a basic structure of three quarks. Their antiparticles, antibaryons, are made from three antiquarks. Mesons are built from a basic structure with one quark and one antiquark, so their antiparticles are mesons too (indeed, in the special case where the quark and antiquark are the same type, the meson and its antiparticle are identical). All then known hadrons could be understood in this way. The omega-minus, whose discovery confirmed the hadron patterns, is a baryon made of three strange quarks.

In the presentation speech of Gell-Mann's Nobel Prize award, which discusses a lot of his earlier work prior to the quark idea, one finds the following statement: "It has not yet been possible to find individual quarks although they have been eagerly looked for. Gell-Mann's idea is none the less of great heuristic value." The Nobel committee, like most physicists in 1969, were very ambivalent about the quark idea, because quarks must be assigned peculiar electric charge values, values that are never seen. Today physicists are convinced of the reality of quarks, and not just their heuristic value.

Helen says: "In the late 1960s many of us did a lot of physics by a method that today seems quite weird. We were secret believers in quarks, we used them to make our calculations simpler, and then we threw them away and wrote about the calculation in an abstract formalism called *current algebra*. We assumed there was a set of currents that could flow inside hadrons. The properties of these currents could be derived from quarks if we assumed that the quarks behaved as if they were free particles in very high-energy interactions. Since we did not yet know that this was indeed the way the QCD theory of strong interactions actually behaved, this seemed an illegitimate assumption. No one had convinced the world that they had seen quarks inside hadrons. So we just said we assumed, the properties of the currents and then the results followed! The quarks were there in the desk drawer versions of the calculations, but erased or much deemphasized in the published versions. Of course they were also there in our conversations, but only provisionally so. We felt we needed better experimental evidence that they existed before we could make them unalterably part of our theory. We also needed further theory developments, developments that happened in the 1970s, before we could go beyond the current algebra and get to quark level calculations that could simply not be done with just the current algebra approach."

The conceptual world of a scientist often has this strange split personality feature. One must commit to new ideas before they are confirmed in order to work on them and move the subject forward. However, one must keep a very clear hierarchy in one's head: this feature of the theory is well established, that one is untested and hence provisional. Physicists often use the term "belief" to describe the provisional state of mind: I know the first set of things, I believe the second are probably right. However, this type of belief is very different from the belief of a true believer—it is a guess or a hypothesis, and it can be changed by contrary evidence. Any good scientist knows very well what in the theory is well established and what is not yet tested.

More recent discoveries have expanded the list of quark types to six quark flavors, plus their six antiparticle flavors. All the more massive quarks are unstable; so then are the particles containing them. Ordinary matter has protons and neutrons, which are formed from only the two lightest quark types. With the introduction of quarks, the stage was set for the development of the modern theory of particles and their interactions. Finally the strong nuclear force could be understood, as an effect of the forces between quarks.

## Why Don't We See the Quarks?

But while the quark proposal solved the puzzle of why so many different hadrons, it came at a price—a new and deeper puzzle appeared. Why had the quarks never been seen?

To give the correct properties for the known particles, the quarks had to be assigned electric charges that were a fraction ($-1/3$ or $+2/3$) of the charge of a proton. Such a charge had never been observed. In fact, to this day, no one has convincingly detected an isolated fractional charge. Apparently the quarks exist, but only inside hadrons. This is an entirely new situation, something made of pieces that one can never take apart to see the pieces in isolation. It is called *confinement*. We now understand it as a property of the forces between quarks.

It is all very well to invent a rule that quarks are confined, but what do we mean by invoking constituents that cannot be separated out? In what sense do the quarks really exist if we cannot see them one at a time? This issue bothered physicists enough that, in the 1960s and even into the early 1970s, the quarks were treated by most physicists as a mathematical fiction, a trick for keeping track of all the particle types, rather than actual physical objects, somehow constrained to exist inside hadrons only.

But, while one cannot produce the quarks in isolation, one can indeed observe them as the substructure of protons and neutrons. The experiments that first did this, carried out in the late 1960s and early 1970s at Stanford Linear Accelerator Center (known as SLAC), were, in a way, a large-scale repeat of the experiment by which Rutherford and his associates Geiger and Marsden (who actually were the ones who did the experiment) first found that atoms contain a small dense nucleus. Geiger and Marsden used alpha particles from a radioactive source and showed that they bounced back in an unexpected way from a thin foil, indicating that they were encountering small dense nuclei within the much bigger size of an atom. Richard Taylor, Henry Kendall, and Jerome Friedman, and their collaborators at SLAC, used high-energy electrons from the then new accelerator, and showed that they bounced back in an unexpected way from the protons and neutrons in a gas target.

Helen says: "At this time I was a little further along in my graduate studies and was working on my thesis at SLAC. Professor James Bjorken was my thesis advisor. He is a theoretical physicist who likes to work closely with experimental data. His suggestions were invaluable for the experimenters. He suggested to them how to organize their data, to show that it gave evidence for substructure in the neutron and proton. One saw that electrons bounced off the protons and neutrons in a way that could be understood if they had small dense constituents but not if they were some kind of uniform blob. I vividly remember his excitement as he showed me the early plots of the data and explained what they meant; this happened sometime in late 1966 or early 1967."

It took more data and a little longer for the rest of the physics community to be convinced that the experiment, as analyzed by Bjorken's methods, was indeed demonstrating the existence of quarks. Only after further developments had supported the quark picture did the community as a whole accept this interpretation of the experiments. Eventually, in 1990, the leaders of the experiment were awarded the Nobel Prize in physics for doing just that. This work is now recognized as the first observation of the quark substructure of protons and neutrons.

Before we proceed, it may be worth reviewing the language of matter and antimatter in terms of quarks. Quarks are matter, antiquarks are antimatter. Quarks and antiquarks, like electrons and positrons, are thought to be fundamental spin-1/2 particles; we have no evidence that suggests they have any substructure, or indeed even any measurable size.

The fermionic (spin-1/2 or spin-3/2) hadrons, known as baryons, have a basic substructure of three quarks. For example, the proton is made of two up quarks and one down quark, while the neutron is made of two down quarks and one up quark. Antibaryons are the corresponding antimatter particles made from three antiquarks. (Note that the difference between a neutron (made of two down quarks with charge $-1/3$ each and one up quark with charge +2/3) and an antineutron (made of two down antiquarks with charge +1/3 and one up antiquark with charge −2/3) is no longer a puzzle.)

Mesons are particles that are neither matter nor antimatter but an equal mixture of both—their basic structure is one quark plus one antiquark. So having classified two types of substance we come upon the third option, a substance that mixes matter and antimatter. But if matter and antimatter can be produced together then nothing forbids their also disappearing together; though of course something else must arise since the laws of conservation of energy, momentum, electric charge, and so on, continue to be true. All mesons are unstable and eventually decay, usually producing lighter mesons. Decays of the lightest mesons produce leptons plus antileptons, or two or more photons. (Another piece of the jargon—spin-1/2 particles which are not subject to the strong nuclear force, such as electrons, muons, and neutrinos, are called *leptons*.)

From the quark point of view, the baryon number, which counts baryons minus antibaryons, is exactly three times the quark number (number of quarks minus antiquarks). Mesons have zero quark number and hence do not contribute to the count. Physicists still use the term baryon number, rather than quark number, as the count of matter minus antimatter particles.

Meson decays can all be understood in terms of the possible decays of more massive quarks to lighter ones, or the production of additional quark–antiquark pairs when the energy to do that is available. In the case of the lightest mesons, a quark and an antiquark disappear, producing in their place, depending on their charge, perhaps a muon and an antimuon neutrino, perhaps some other lepton–antilepton pair, or perhaps a couple of photons.

Each type of meson has a basic quark substructure. For example, a positively charged pion has a basic structure of one up quark and one down antiquark. (Physicists are once again sloppy about the language here; they might also call the antiparticle of a down quark an anti–down quark.) The positively charged pion has an antiparticle, a negatively charged pion. That particle is made from a down quark and an up antiquark. Equally well, we can say that the antiparticle of a negatively charged pion is a positively charged pion.

Our earlier statement, that the K-zero ($K^0$) and anti-K-zero ($\overline{K}^0$) are antiparticles (or CP conjugates) of each other, obtains a simple interpretation in the quark picture: the former has the structure of a down quark and strange antiquark, the latter a down antiquark and a strange quark. (The down and strange quarks carry the same charge, $-1/3$, but differ in mass.) The concept of strangeness and strange particles also becomes simple at the quark level: it is the number of strange antiquarks minus the number of strange quarks. Strangeness is thus a label, much like electric charge, that is carried by one type of quark and not by the others. (By historically defined convention, which predates the quark picture, a strange quark has strangeness $-1$.)

Later, additional mesons that contain charm or bottom (anti)quarks will play an important role in our story. D-mesons have a single charm quark or antiquark. The neutral D-mesons are D-zero ($D^0$) and anti-D-zero ($\overline{D}^0$), which are antiparticles of each other, with the structure of a charm quark and an up antiquark ($c\bar{u}$) for the former and a charm antiquark and up quark ($u\bar{c}$) for the latter. The charged D-mesons are the CP-conjugate pair of D-plus ($D^+ = c\bar{d}$) and D-minus ($D^- = d\bar{c}$). Similarly, there is a quartet of B-mesons: B-zero ($B^0 = d\bar{b}$), anti-B-zero ($\overline{B}^0 = b\bar{d}$), B-plus ($B^+ = u\bar{b}$), and B-minus ($B^- = b\bar{u}$), and so on.

Some neutral mesons are matter–antimatter mirrors of themselves; for example, a phi-meson is made of a strange quark and a matching (anti) strange antiquark. If each of these is turned into its antiparticle we still have a phi-meson. So the antiparticle of a phi-meson is a phi-meson. Note that this is also true for the uncharged fundamental particles with integer spin—for which we use the generic term bosons—the antiparticle of a photon is a photon. (Physicists prefer this way of saying that these neutral bosons have no distinct antiparticle partner because it allows them to consider interactions and the mirror interactions in which each particle is replaced by its matching antiparticle without having to state special cases when such self-conjugate neutral mesons are involved.)

Note, however, that while mesons are neither matter nor antimatter, they contain both matter and antimatter and thus have all the interactions of ordinary matter-type particles. This means that these particles are not the stuff that makes up the mysterious dark matter. Astronomers and cosmologists use the terminology *baryonic matter* for the stuff that makes up stars and, indeed, it is mostly baryons, protons and neutrons, but it also includes mesons.

## What about Dark Matter?

With all the new ideas and new particle words, it is easy to confuse one new concept with another. So let us be quite clear—dark matter is not, and could not be, the missing antimatter. The massive regions of dark matter that are found to extend beyond the visible parts of galaxies have to be made from some entirely different type of particles; they contain neither quarks nor antiquarks. We can model the growth of galaxies, and to get them to form with a dark matter halo requires very different properties for dark matter than for ordinary matter, in particular it requires that the dark matter does not have strong or electromagnetic interactions. So it cannot be made from quarks or antiquarks because they have both types, and it cannot be made from charged leptons because these have electric charges.

Furthermore, the neutral leptons, the neutrinos, have such small masses that, when they are used to model dark matter, they do not work either, they give a completely wrong pattern of formation of galaxy clusters. (They form a part of the dark matter in the Universe, but not the major part.) So dark matter is another mystery; it tells us that our theory of quarks and leptons, matter and antimatter, baryons, antibaryons, and mesons, is still not a complete story of all the things that have mass.

Most probably the dark matter is made from some other type of particle. Observations of galaxies show that it is there, but as yet it is not part of our standard particle theory. Whatever they may be, dark matter particles are very hard to produce. Even though we are swimming in a sea of them, the dark matter halo of our galaxy, they are hard to detect in any laboratory. Some ideas for extending the theory introduce possible dark matter particles. Each idea suggests certain types of experiments to search for the particles it proposes.

There are now experiments searching for two possible types of dark matter: one set of experiments looks for particles called neutralinos (or more generically WIMPs—weakly interacting massive particles) and the other for a type of particle called axions that have much smaller masses. Since these particles, the axions, arise from a theory that Helen proposed with her collaborator Roberto Peccei for a quite different reason, to explain the CP symmetry of the strong interactions, she is, of course, a fan of that answer. Yossi, on the other hand, likes supersymmetry, the extension of the theory that suggests the neutralinos, so he bets on that side. If the dark matter that is in the halo of our own galaxy is made of either of these types of particles, we should know it before too much longer. Neither axions nor neutralinos are part of the Standard Model, but they are both suggested by extensions of the theory first introduced for other reasons, not just added to fit the dark matter.

## The Missing Charm, the Surprising Tau

By now we have mentioned most of the known particle types, but we also need to say a little about the patterns of their interactions. Our present understanding of fundamental interactions among elementary particles is encoded in a set of mathematical equations that are called the *Standard Model*: standard because it has survived so many tests it is the basis of all our understanding of particles, and model because it started out as one model among many. By now this term is a misnomer. The "model" is truly a well-established theory. Furthermore, the word "standard" is a dull name for a theory of this beauty and scope. It has a very elegant (and powerfully predictive) symmetry principle as its starting point.

Later we describe the Standard Model in its recent form. This current picture developed over time. Here we tell you some crucial steps in the history, which reveal important insights that are related to matter and antimatter. Their significance to baryogenesis will become clear later.

When the Standard Model was first formulated, that is, in the early 1970s, it included the four leptons and four quarks that were known at that time. These were conveniently divided into two so-called *generations*, each consisting of a quartet of particles of different electric charges: the up quark (charge +2/3), down quark (charge −1/3), electron (charge −1), and electron-neutrino (charge 0) constitute the first generation, while the charm quark (charge +2/3), strange quark (charge −1/3), muon (charge −1), and muon-neutrino (charge 0) constitute the second.

Indeed not even all four quarks had been observed. All ordinary matter is composed of particles of the first (lightest) generation. But experiments had revealed the existence of other types of matter, short lived, but produced in cosmic rays and in accelerator experiments. These required the third quark, the strange quark, and also included the muon and muon-neutrino. But particles containing the fourth quark, the charm quark, needed to complete the second generation, had not been observed. Its existence was a prediction of the Standard Model.

If you tried to write a Standard Model type of theory with only three quarks, it would give wrong predictions. For example, it would predict that the neutral K-long meson should decay to a muon–antimuon pair at a rate much larger than observed. The fact that this problem could be fixed by adding a fourth quark was seen, and hence such a quark was predicted, based on symmetry arguments. In the early 1970s, many physicists were loath to accept the extension of the quark model with such symmetry arguments alone to justify the prediction of a new quark type. The idea of postulating the existence of the many new particle types that could be formed using such a quark, just to avoid particular weak interaction effects, seemed far fetched. So they looked for other ways out of the problem.

Indeed at that time many physicists had not yet accepted the idea that the first three quarks were real objects; after all they had not been observed in isolation. Evidence for their existence as the substructure of protons and neutrons was indeed accumulating, but its interpretation was not, as yet, widely understood. As is usual with a good new idea, there developed a subculture within physics of people who took the quark idea seriously, and whose work was directed toward understanding the implications of this idea in more detail.

Other competing ideas had their own subcultures too, for example the idea that there is not any really fundamental level but that every hadron can be equally well viewed either as a fundamental particle or as a composite of other hadrons. In that viewpoint, a proton could be a composite of a pion ($\pi^+$) and a neutron, but a neutron could equally well be a composite of a proton and a pion (of a different charge, $\pi^-$). Worse yet, a pion had to be viewed as being a composite of a proton and an antineutron (or neutron and antiproton if it had negative charge). This idea, variously called nuclear democracy and the bootstrap (for pulling oneself up by one's own bootstraps) had many proponents, and indeed it led to some insights that have outlasted the idea itself. But the number of adherents to this camp shrank rapidly after dramatic successes of the quark theory in the early 1970s paved the way for the modern Standard Model.

Helen recounts: "I remember being in my office at Harvard University when Sheldon Glashow, who was one of those who made the suggestion that charm quarks were needed, stopped by to see who wanted to go to lunch. He listened to our conversation for a moment and then said—'Do you mean to tell me that I'm in a room full of people all of whom believe in charm?' Indeed, everyone in the room took seriously not just the quark idea itself, but also his suggestion of a fourth quark type. We had been talking about what evidence for a fourth quark might be seen in electron-positron collision experiments. There were so many of us, five in fact (Tom Appelquist, Howard Georgi, Alvaro de Rujula, David Politzer, and me). Shelley could not believe that we all thought alike on this topic. That was in May 1974. (In fact there was a sixth person in the room, my infant son James, but, since he was asleep and anyway had no opinion on the matter, Shelley did not include him when saying 'all.') The discovery in November 1974 of a meson, now known as the J-psi ($J/\psi$) particle, containing the charm quark along with its antiquark, was the first major victory for the theory. By November 1975 you couldn't find a room big enough for all the physicists who took the idea of charm, and with it, the reality of quarks, seriously."

The discovery of these new particles in November 1974 was so dramatic that particle physicists call it the *November Revolution*. Within hours of the seminars that announced it, physicists all over the world were calling one another to share the news and talk about their favorite interpretations of the data. The log book from the experiment at SLAC shows that the people running the experiment the night before the announcement were excited too. The log book says "the events came pouring in!!!" when the energy of the colliding electron and positron was set just right.

Definitely something new was being seen. There was champagne flowing in the counting room before morning. (Today's safety rules would forbid that, but in 1974 they were not as explicit.)

Helen again: "I was teaching a junior level course on electromagnetism at Harvard that term. I did not give the lecture I had prepared the night before. When I arrived at Harvard that morning Tom Appelquist told me about the experiment. We immediately jumped to its interpretation in terms of charm; it was exactly the signal we had been talking about that had prompted Shelley's comment a few months earlier. My students that day heard about the latest excitement in particle physics and this interpretation of the data, not the theory of electromagnetism. In all my years of teaching this is the only time I ever preempted the course topic for the news of the day—I just could not think or talk about anything else that day."

Remarkably, two quite different experiments announced the discovery of this same particle in seminars on the same day and at the same place, one after the other. One experiment, led by Samuel Ting, took place at Brookhaven National Laboratory, on Long Island; the other, led by Burton Richter, at Stanford Linear Accelerator Center in California. Ting happened to be at Stanford the day that the discovery was made there, heard some word of it and quickly called his colleagues to send him their data, which he knew showed the same particle. The two groups gave back-to-back presentations at SLAC to announce their discovery to the physics community.

In fact Ting's group had made the discovery some months earlier, but until that day Ting had ensured that this result was kept secret; he wanted to carefully check many features of this discovery before he announced it. The fact that Ting could achieve this level of secrecy shows the strong hand with which he ran his group. Helen remembers: "After my husband and I got our Ph.D.'s, we accepted positions at the Deutches Elektronen Synchrotron in Hamburg Germany. Dan, then an experimental physicist, joined Sam Ting's group. There are many stories I could tell of his experiences which illustrate Sam's style of management of this group. Here's just one—one evening, when we had gone into Hamburg to the Opera, we got back to our parked car to find a note on it which said— 'What are you doing taking the evening off? You should be working. Sam.' Dan turned white and began to get very upset, but then realized the note was not Sam's handwriting. One of his other colleagues had played a joke on him. However, his first reaction to the note was no joke; it was complete belief that Sam might do this and mean it. That belief told a story." When Ting moved on to experiments at Brookhaven he apparently took the same management style with him.

However, when Ting heard the buzz that morning at SLAC, he realized he was about to be scooped if he waited any longer; he soon decided enough checking had been done. The dual announcement accounts for the particle's strange dual name (and the shared Nobel Prize). The tradition is that the discoverers get to name the particle. But the two groups had chosen different names, and neither would yield the honor to the other, so the east coast *J* and the west coast ψ both persist; its official name is the awkward combination *J*/ψ. Very similar but slightly more massive particles discovered soon after at SLAC have only west coast names, they are called psi-prime, psi-double-prime, and so on.

Charm was the explanation immediately suggested by many theorists (including those from that room at Harvard) for this new particle. It could be a meson made from a charm quark and a matching charm antiquark. But many other ideas too were proposed to explain the events. Each such proposal made different predictions about what would happen at slightly higher energies. It took over a year for the experiments to check that the patterns predicted by the theory for the charm quark were indeed present in the data. But eventually even the most skeptical physicists had to agree that charm was the best explanation. The two-generation Standard Model had won its first major victory. The name "Standard Model" began to be used.

Part of the reason for the slow confirmation that the charm quark is the right explanation for the J-psi particle was an extraordinary coincidence. It turned out that not one but two totally new types of matter (and matching new antimatter) were being produced in the experiment at SLAC at almost the same energy. The second discovery, made by Martin Perl (1927– ; Nobel Prize 1995) and his collaborators, was that there is a third charged lepton, the tau. This could only fit the picture as the first particle of a third generation of matter.

Martin Perl follows his own drummer. In his Nobel lecture he tells us: "I reflected that it would be most useful for me to pursue experiments on charged leptons, experiments which might clarify the nature of leptons, or explain the electron-muon problem. This is a research strategy I have followed quite a few times in my life. I stay away from lines of research where many people are working . . . I find it more comfortable to work in uncrowded areas of physics." What Perl and his team looked for, while most others were focused on hadron events to understand charm, were events where (after an electron and positron collided and annihilated) the only particles seen in the detector were one electron or positron and an oppositely charged muon. Furthermore, the tracks of these two oppositely charged particles showed that they were not traveling back to back. These electron-muon events could only be interpreted by positing that a new additional heavy lepton, the tau, had been produced, together with its oppositely charged antiparticle, and then both had decayed, one to a muon (or antimuon) and two (unseen) neutrinos, and the other to a positron (or an electron) and two neutrinos. The unseen neutrinos explain the peculiarity that the only two tracks in the event are not back to back, which would violate conservation of momentum if these were the only particles produced. These peculiar events showed that the tau must exist. There was no other process that could cause them at the observed rate.

It took some time for Perl to collect a large enough sample of his electron-muon events to convince the rest of the team he really had something noteworthy. Furthermore, one had to figure out something new about tau decays. The electron-muon events are not the only possible way for a pair of tau particles to decay. In its decay, a tau lepton emits a W-boson and becomes a tau-neutrino. But the tau is heavy enough that the $W^{\pm}$ from this transition can produce not only a charged lepton and a neutrino. It could, and indeed more often does, produce a quark and antiquark (one up type and one down type). Hence the tau, although it is a lepton, can decay to a set of hadrons (plus an unseen neutrino). This explains the time it took to confirm that charm quarks were a good explanation for the psi-type particles. The predictions for the pattern of events from decays of mesons containing a charm quark (or antiquark) together with a lighter antiquark (or quark) matched the data only after all the events that were due to tau production had been removed from the accounting. Or, to say it another way, the observed patterns could be explained only by adding to the processes that occurred at lower energy the events expected from the production and decay of both charm–anticharm and tau–antitau pairs.

Few physicists expected yet another generation; two seemed peculiar enough. A few had been bold enough to speculate about three generations, but the majority of particle physicists were still struggling to accept the fourth quark, two more seemed quite superfluous. But all four members of that third generation have since been found: the top and bottom quarks, and the tau and tau-neutrino. The pattern of charges of the third-generation particles is identical to that of the first- and second-generation particles, but their masses are heavier (with the possible exception of the neutrino).

At present we have plausible reasons to believe that there are only these three generations of particles and there is no fourth, heavier one. These reasons are not completely watertight. Some day you may indeed hear that there are more fundamental particles than these three generations of matter particles and their matching three generations of antimatter. But for now the Standard Model seems complete with just three generations. The fact there are at least three copies of matter fermions, that is three generations, turns out to be important in the way differences between matter and antimatter enter the theory.

## The Standard Model: Particles and Interactions

The two basic ingredients of the Standard Model are the symmetries that define it and a list of elementary particles. An amazing feature of the Standard Model (and of most hypothetical extensions of it) is that the symmetries determine (almost) entirely the various interactions that can take place and their features. We here describe the interactions. We will come back to the relation between symmetries and interactions in later chapters.

Let us start by simply listing the interactions and particles that are described by the Standard Model. The Standard Model includes three types of interactions: the strong interaction, the electromagnetic interaction, and the weak interaction. (The fourth known type of interaction, that is, the gravitational interaction, is left out of the Standard Model. Indeed, it plays no role in experiments of high-energy physics, because it is much much weaker than all other interactions at the energies at which these experiments are conducted.) In the Standard Model there are twelve elementary, spin-1/2, matter particles: six quark types (top, charm, up, bottom, strange, down) and six lepton types (tau, muon, electron, tau-neutrino, muon-neutrino, and electron-neutrino). Since these particles are fermions, they all obey Dirac-like equations and thus all have antiparticles. Now let us make ourselves more acquainted with these interactions and particles.

Particles are subject to a certain type of interaction if they carry the appropriate charge. Of the twelve elementary particles listed above, the first six (top, charm, up, bottom, strange, and down), known collectively as quarks, are subject to the strong interaction because they carry the appropriate charge, which physicists call *color* charge, although it has nothing to do with the usual meaning of the word color. The color charge of quarks comes in three varieties, red, blue, and green, if you want. (In analogy to real colors, the combination of three quarks of different colors is white, that is, color-neutral and consequently insensitive to the strong force.) All quarks carry also an electric charge, and are therefore subject also to the electromagnetic interaction: up-type quarks (top, charm, and up) have charge +2/3, while down-type quarks (bottom, strange, and down) have charge −1/3.

The other six elementary particles (tau, muon, electron, tau-neutrino, muon-neutrino, and electron-neutrino) are collectively known as leptons; they have no color charge and are insensitive to the strong force. The charged leptons—electron, muon, and tau—have charge −1, and thus interact electromagnetically. The neutrinos have no electric charge and therefore are subject to neither the strong force nor the electromagnetic one.

If we were to classify the spin-1/2 particles according to their color and electric charges, we would find four classes of particles: uplike quarks (colored, charge +2/3), downlike quarks (colored, charge −1/3), charged leptons (colorless, charge −1), and neutrinos (colorless, neutral). As we explained before, a quartet of four particles, one of each type—uplike quark, downlike quark, charged lepton, and neutrino—is called a generation. Thus we say that the Standard Model has three generations; and we have found that there are three particles for each of these classes. Particles in each generation have different masses from the same-color same-charge particles in other generations.

The term "generation" suggests some form of parent-child relationship, between the particles of one generation and the next. Decays of second-generation particles often yield first-generation particles. It seems we labeled the generations backward. But the relationships suggested by particle decay patterns are, in the end, too complex to be thought of in this simple parent–child way. So, while the name stuck, do not be caught up trying to understand its implications. It is simply a term for a set of four particles, two quarks, one of each charge, and two leptons, likewise one of each charge.

When we say that a particle is subject to a certain interaction, we mean that it is capable of exchanging certain spin-1 particles, known as force carriers. Thus, an electron and a positron interact with each other by exchanging photons, the carriers of the electromagnetic force. Three quarks are confined into a proton by exchanging gluons, the strong force carriers, with one another. Nuclear beta decay occurs when a quark emits a W-boson, a weak force carrier. The W-boson then produces an electron and an antineutrino.

The electromagnetic force carrier is known as the photon. The photon is a spin-1, massless particle. Its masslessness is responsible to the fact that the electromagnetic interaction is long range. Multiple photons make up electromagnetic waves, and thus all forms of electromagnetic radiation (radio waves, light, X-rays, and gamma rays, etc.) can be thought of in terms of photons.

The spin-1 strong force carriers are known as gluons, because they glue the quarks together into hadrons. They come with eight possible color charges. The fact that they carry these charges of strong interactions makes them behave very differently from photons, which carry no electric charge of their own: Gluons can emit and absorb other gluons. Gluons are also, technically speaking, massless particles. The strong interaction, however, confines colored particles (both quarks and gluons) into colorless combinations (hadrons). Since hadrons have a very small size we never observe long-range strong interactions or directly measure the mass of a gluon. Thus the gluon's zero mass is a formal (mathematical) rather than a physical (observable) property.

Weak interactions have three spin-1 force carriers, the two charged W-bosons, W-plus ($W^+$) and W-minus ($W^-$), and the electrically neutral Z-boson, Z-zero ($Z^0$). The superscript on the weak force carrier names—plus, minus, or zero—corresponds to the electric charge carried by them. Unlike the other Standard Model force carriers, the $W^+$, $W^-$, and $Z^0$ are massive, with masses that are about a hundred times that of the proton. This feature makes the weak force short range, and this, rather than a small coupling strength, accounts for the fact that its effects are rare compared to electromagnetic processes. In a sense, the term weak for the weak interactions is another misnomer. The weak interactions are weak only at relatively long distances. Surprisingly, for short enough distances, the strengths of the three types of interaction are quite comparable.

It is worth taking a moment to say more about the discovery of the Z-boson. This particle was predicted by the Standard Model theory but its effects had never been seen, because, in most circumstances, they are hidden by much larger electromagnetic effects, from photon exchange. But once the predictions were made, careful experiments could be designed to look for the Z-boson, by observing effects such as parity violation that could not be due to photons, and, eventually, to produce Z-bosons and observe their decays. Each step along the way matched the theory and gave us stronger and stronger evidence for the existence of these particles and the interactions that they mediate.

When we listed the particles, we did not say which are subject to the weak interactions and which are not. This question is somewhat more complicated, as here we have to talk about left-handed or right-handed states, rather than about particles. Massive particles have, unavoidably, both left-handed and right-handed components, so such a division to states does not seem to make sense. We will later explain this issue, but for now let us simply think of all matter particles as massless, and thus their left-handed and right-handed components have an independent existence. All left-handed matter particles (and right-handed antiparticles) carry weak charge. All right-handed matter particles (and left-handed antiparticles) carry no weak charge and therefore are unable to exchange the weak force carriers. This is the way that parity violation by the weak interaction is encoded into the Standard Model. Left-handed particles carry a different weak charge from their mirror-image partners, the right-handed particles. In contrast, the color of left-handed and right-handed particles is the same, as is their electric charge. Consequently, strong and electromagnetic interactions are the same in our universe and in its mirror image.

For one type of particle, parity violation in the Standard Model could go even further: for neutrinos, the Standard Model can include only left-handed ones, with no right-handed neutrinos. Likewise it would then have only right-handed antineutrinos and no left-handed ones. This does not affect the parity invariance of the equations that describe strong and electromagnetic interactions: since neutrinos carry neither color nor electromagnetic charge, they simply do not appear in these equations. However, now that we know that neutrinos are not massless, we can no longer write a theory for them without introducing right-handed neutrinos and left-handed antineutrinos. These things must exist. Even though the Standard Model interactions did not require us to include them in the theory, the observation that neutrinos are massive forces us to add them in.

# 11

# ENERGY RULES

## Stored Energy, Forces, and Energy Conservation

The mathematical language used in modern particle physics to understand the fundamental interactions and to write the Standard Model is that of *quantum field theory*. This is a general mathematical formalism, within which one can write many possible physical theories. But only one of them will describe the world we live in. Our job as particle physicists is to decipher what that one might be. The field theory formalism supersedes but builds on the Dirac equation. In this chapter, we will explain how physicists think about energy, and what they mean by the word fields. We will try to give you a glimpse of what quantum field theories are all about, without resorting to mathematics. This will then allow us to describe the modern understanding of symmetries (and their breaking) in general and CP symmetry (and its violation) in particular.

To our senses, forces are obvious, but energy is a rather abstract quantity. In physics, it turns out that energy is the primary idea and forces are a secondary concept. Particle physicists think about forces as a consequence of energy patterns. The source of all forces is the distribution of energy in a system. Forces are the signal that stored energy can change when components of the system move around.

Any force on an object pushes it in a particular direction. The force on any object due to a given type of interaction is always in the direction such that motion of the object in that direction would decrease the stored energy associated with that type of interaction. Thus the force of gravity pulls any two massive objects closer together.

Of course this does not always mean that the object moves in the direction of a particular force on it. It may already be moving in another direction, so the force changes it motion, deflecting it or slowing it down. Alternatively there may be other forces, due to other interactions, that either completely or partially cancel out the first force.

You have met the idea of stored energy even if you never took a physics course. You recognize that food or fuel contains stored energy. You know that a compressed spring contains stored energy too, and that it pushes in the direction that will reduce its compression and thus its stored energy. Another example is better for our purposes. Think of the water in a high-walled reservoir. You know that if the dam is breached the water will do terrible damage to the valley below as it rages down it. Or, in more controlled circumstances, that it will pour down a penstock and turn a turbine to produce electricity. So the water in the dam has stored energy or, to use a more technical name, gravitational potential energy. Now think of a rock held in the air and imagine letting it go—it too must have had stored energy because it starts to move, and motion is certainly a form of energy. The law of conservation of energy tells us that increase in motion energy must come at the cost of a decrease in stored energy, just as it does when your car burns gasoline to make it move.

Notice that these examples convert stored energy to motion energy, but the reverse process is also possible, once a particle is moving. The simplest example is a weight bobbing up and down on the end of a spring. At the top of its motion the weight is still but gravitational stored energy and energy stored in the compressed spring can both be reduced if it moves down, so it feels a net force in that direction and begins to accelerate downward. In the middle of the motion there is a point where gravity pulls down and the spring pulls up with equal force, so the weight at that point feels no force and is not gaining or losing any energy. However, it is moving, so it goes right on by that point and gets to where the force of the spring pulling up is greater than the force of gravity pulling down, so now it slows down, its motion energy decreases, and the energy stored in the spring increases, until finally the weight stops (for an instant only) at the bottom of its motion. But at this instant there is an upward force on it, because it can reduce the energy stored in the spring more than it increases the energy stored in its gravitational interaction with the Earth by moving up. So up it goes again, and, if we ignore any dissipative effects, it eventually gets back to the top and repeats the whole cycle. Energy goes back and forth between stored energy and motion energy over and over again as it bounces.

Interplay of different forces occurs whenever motion of an object can rearrange stored energy. If there is more than one reservoir of stored energy in the problem there can be forces but no motion. For example you are probably sitting on a chair. So let us consider the system that includes you, the Earth, and the chair. The force of gravity pulls both you and the chair toward the center of the Earth because that would minimize the gravitational stored energy. So why don't you and your chair both fall there?

What nonsense you say—I can't fall through solid earth. But what makes the earth solid? The earth pushes up on the chair, counterbalancing gravity pulling down, and the chair in turn pushes up on you. Where do these upward forces come from? They arise because of electromagnetic forces between atoms that form molecules, metals and crystals, soil and rocks, the frame and cushion of your chair. In isolation, each lump of material arranges its atoms so as to minimize as best it can the electromagnetic energy stored within it. When your chair pushes down on the floor it increases the local stored electromagnetic energy in the floor a little, just as happens when you compress a spring. But the gravitational stored energy is reduced by the fact that the chair moves down a little and you move down even more by compressing the cushion.

The floor holds the chair up, and the chair holds you up only if the total stored energy, electromagnetic plus gravitational, of you, the chair, and the Earth, is minimized before the chair or the floor finds a way to rearrange itself by stretching out so that the atoms of the solid material part company and you fall through. Put a heavy enough object on your chair and you will see what happens when the gravitational stored energy reduction that can be achieved by lowering the heavy object is bigger than any increase in electromagnetic potential energy that can be made in the structure of the chair—some part of the chair crumples or breaks and the chair collapses (figure 11.1). But when you are comfortably sitting still on your chair, the downward force of gravity on you is exactly balanced by the upward force on you from the material of the chair. Moving either

Fig. 11.1 Put a heavy enough object on your chair and you will see what happens.

up or down would increase the total stored energy compared to that where you sit compressing your cushion and bending your floor a little too, though that effect is probably too small to notice without careful measurement.

## Force Fields Permeating Space

Just as gravitational interactions lead to gravitational potential energy stored between two massive objects, such as a rock and the Earth, so all types of interaction store energy. To keep track of how the energy is distributed in space and how this leads to forces, we need the concept of a force field. You may think of this as a science fiction concept, but think again. No doubt you know about magnetic fields and have seen a compass orient itself in the Earth's magnetic field. You may possibly even have heard of electric fields. You may never have heard of a gravitational field, though you certainly live immersed in one and feel its effects all the time. These are all force fields. But what are these force fields and what do they have to do with stored energy?

Most people have played with magnets and seen that they exert forces on one another even when they are not touching. The compass is just one example of this. We can describe the effect of one magnet on any other in terms of magnetic fields. Similarly we can encode the effects of one electric charge on another in terms of electric fields. But what is a magnetic or electric field? Are they real, or just a mathematical fiction? A static unchanging field is a thing we cannot see. But light, the only thing we actually do see, is nothing more than a traveling wave of changing electric and magnetic fields. So when these fields vary at the right frequency we do see them.

Electric and magnetic fields contain energy. That is why we can see light, because it transfers energy to our retina. All the energy that comes to us across space from the sun is carried by a moving wave of electric and magnetic fields. All that moves or changes in such a wave is the strength and direction of the electric and magnetic fields at each point in space. Clearly these fields are not just a mathematical fiction. The energy we get from the sun is very real.

When the fields are static (that is, unchanging) we cannot usually see or feel them, though a strong enough electric field can make your skin tingle and your hair stand on end. But even when we cannot feel them, we can detect them with instruments designed to be more sensitive than our natural senses. We can measure what force a test charged particle or magnet is subject to when placed in the field. So charges and magnets are not just sources for these fields, they are sensors for them too.

Electric and magnetic fields are a consequence of the electric charges and magnets and their motion. At first electric and magnetic fields look like two separate and unrelated effects. But experiments show that a moving charge (or an electric current) also creates a magnetic field. This immediately tells you that electricity and magnetism are closely related effects. An observer in a frame of reference moving along with the charge sees a static electric field around the charge. But for an observer for whom the charge is moving, there are both a time-varying electric field that depends on the distance between the observer and the charge and a time-varying magnetic field. What is electric and what is magnetic depends on the frame of reference—clearly we cannot treat these as entirely separate things. Thus we used the term *electromagnetic* for the combined electric and magnetic phenomena.

The electromagnetic interaction is a name for all the effects of the interplay of electromagnetic fields and electric and magnetic charges (a magnetic charge is also called a magnetic pole, and they come in north and south pairings). We have not described all the effects of electromagnetism, but just enough to try to give you a sense of the reality of these space-filling fields, and their consequences.

The combined understanding of electricity and magnetism in terms of electric and magnetic fields was a major triumph of nineteenth-century physics. James Clark Maxwell brought it all together in 1873 with his formulation of the equations we now call Maxwell's equations. Maxwell's equations encode the rules for the patterns of fields produced by charged particles and magnets and predict the forces exerted by these fields on other charged particles and magnets.

The important idea we need is that a field is a quantity that can take a value (a size, and possibly a direction in which it points) at any point in space and time, so, mathematically speaking we say it is a function of space and time. Of course, there are times and places where a particular field is not present; in our formalism this simply means that the value of that field then and there is zero.


## Field Theory and the Energy Function

All the effects that we have been describing can be codified by writing the equations for how the energy is stored in the fields, and how it depends on charges, their locations, and their motions. These equations comprise the mathematical physics formalism known as quantum field theory. The expression that describes the energy density in terms of the fields, their strengths, and their rate of spatial variation at each point in space is the defining quantity for a field theory. It encodes all the forces and interactions and predicts the behavior of particles.

We will call this quantity the *energy function*, though its proper name is the *Hamiltonian*, named for Sir William Rowan Hamilton, the person who first recognized (in 1835) the power of this formulation of physics. This, as you can see, was well before quantum physics, or even electromagnetism, was understood. Hamilton's contribution was to recognize, in the context of ordinary everyday mechanical systems, the power of the concept of energy and the possibility of predicting how a system will act once all factors that affect its energy are accounted for. Not only that, he found a very compact way to represent this information mathematically.

To be precise, the quantity used to define field theories is actually a closely related function, named the *Lagrangian* after the great French mathematical physicist Joseph-Louis Lagrange, who devised the alternate formalism at about the same time as Hamilton's. The difference between the Hamiltonian and the Lagrangian is technical, the key concept for both is that the behavior of a system depends solely on the way the kinetic and potential energies change as elements of the system change, so we will continue to use the term energy function without making any distinction as to which of these functions we are talking about. The difference is too technical to matter here.

In fact the idea of codifying the energy in terms of fields, which was developed first for electromagnetism, works so well that in particle physics theories we take it a step further. To be honest, the field that appears in the field theory of electromagnetism, known as quantum electrodynamics, or QED, is neither the electric nor the magnetic field but a separate and more abstract function of space and time which contains information that specifies both the physical electric and magnetic fields. It is this more abstract field function (originally called the electromagnetic potential function) that represents the photons as well as any static electric and magnetic fields in the field theory formalism. From now on we will call this the photon field.

Thus quantum theory introduced the idea that particles can have wave-like behavior. Einstein explained the photoelectric effect (the work for which he was awarded a Nobel Prize) when he recognized that it showed that quantum mechanics requires that light, which we have just described as a wave-like spatially changing field, an electromagnetic wave, can have particle-like behavior—arriving in discrete chunks with well-defined energy (photons). This suggests that the distinction between matter and radiation or force fields is somewhat artificial. This is called *particle-wave duality*. It was one of the biggest surprises of quantum physics. Indeed, certain experiments could only be understood if you assigned electrons wave-like behavior, so that there are interference effects involving particles.

Physicists took a while to get used to this odd situation; historically particles and electromagnetic radiation seemed to be entirely different kinds of things. But eventually they came to understand that the only approach that gave results consistent with all observations was to use a single formalism that encompasses both the matter and the force particles by introducing a field associated with each type. Quantum field theory is the modern descendent of both quantum mechanics and the classical field theory of electromagnetism. It takes particle-wave duality to the logical conclusion. In it, we introduce both force fields and matter fields. Each type of field gives both particle-like and wave-like effects. The total energy in any region can then be expressed in terms of the values of all the fields in that region.

Once there is a field there can be an excitation that is a little lump of field traveling along—a particle—and similarly there can be excitations that are waves of changing value of the field, traveling through space. In fact there can be any number of such excitations. The words *quantum excitation* simply mean the smallest lumplike pulse of this field that occurs—this is the single particle. Each particle type has a definite mass and a definite spin. All these properties can be specified by choosing the mathematical type of each field, fixing the form of the energy function in terms of these fields, and setting the values of the numerical parameters or constants that appear in that function.

Thus a quark is a quantum excitation of a quark field, just as a photon is a quantum excitation of the photon (electromagnetic) fields. There is a particle type for each field and a field type for each particle. The distinction between what we call particles (matter and antimatter) and what we call force fields (and radiation) is based on our everyday experience of matter and electromagnetic radiation. It relates to the way the mathematics of the theory works too. In the Standard Model, all the fundamental matter-type particles and their antimatter partners have spin 1/2, that is, $1/2\ h$ units of internal angular momentum per particle. However, all the particles associated with force fields have spin 1. The spin requires certain mathematical properties of the associated fields. These properties in turn restrict the allowed terms involving the fields that can appear in the energy function.

Conversely, one can say that the properties of the particles are defined once one chooses what type of field represents them (spin 1/2 or 1 or 0) and chooses the function that defines the energy in terms of the fields. The key part of writing down such a theory is to express how the energy of the system depends on products of the values of the various types of fields, or on their rate of spatial change at each point in space. The parameters that appear in this expression then define the strength of the various interactions and the masses of the various particles.

The energy function is a mathematical expression written in terms of the fields. The equations that predict how the system behaves, both the particle-like and the wave-like phenomena associated with each type of field, can be derived once the energy function is defined. So all the laws of physics are written once the energy function is completely defined.

We say that one field is *coupled* to another if there is any term in the energy function that involves both types of fields. Such a term allows energy stored in one field to be transmitted into energy stored in the other. The *coupling strength* is a number that says how big the effect of such a term in the energy function is. The larger the number, the more likely that energy is transferred from one field type to the other.

What a remarkable shorthand we have found! It takes only a couple of lines to write the energy function (or rather the Lagrangian) for the Standard Model and yet all of the physics we know is subsumed in it—at least in principle. In practice we can do the calculations directly from that starting point only for a very limited and special set of circumstances. But even that limited set provides thousands of tests that the theory has survived with flying colors.

You may have noticed again that there is a certain appearance of circularity in our logic. We can say we write the field theory to encode the observed properties. Alternatively we say that the field theory predicts those properties. But if we just encoded what we already knew, in what way is the theory predictive? In actual fact it goes a little each way. The formalism can be quite restrictive. Once we use it to encode certain observed properties it then automatically makes some significant further predictions.

Tests of those predictions then force additions or modifications to our field theory. The process of building a theory is that of finding what options are available to encode all the known facts, and then of making successive refinements or additions to the theory when its predictions do not match the tests. If the process never ended, if we just could keep adding term after term to our field theory, we would not have a theory with any definitive predictions.

Surprisingly enough that is not what happens. Once we have enumerated all the particle types and symmetries in our theory, various technical limitations restrict us to relatively few terms in the theory, which means there are relatively few adjustable parameters. These parameters are the numerical constants that multiply each possible product of fields in the energy function. We can determine the sizes of these parameters by matching the predictions of the theory to some set of observations. Then all further processes involving the same set of particles can (at least in principle) be calculated; they are predicted by the theory.

The Standard Model example of a field theory was first written down, in its complete modern form, around 1976. Remarkably, as far as we can tell from particle physics experiments to date, it is almost all we need. Only one set of additions have had to be made, because in the late 1990s and early 2000s experiments showed that neutrinos have tiny masses. In the 1976 version of the theory, physicists chose to set those masses to zero, since that was consistent with observation up to that time, and simplifies the theory somewhat.

# 12
# *SYMMETRY RULES*

## Symmetries as Answers to the Question "Why?"

Like the child who refuses to accept the parent's answer "because I say so," so the particle physicists continue to seek deeper reasons for the rules they have discovered. A major role in these answers in modern theories of particle physics is played by symmetries. Remember that, in the physicists' language, the term *symmetry* refers to an invariance of the equations that describe a physical system. The fact that a symmetry and an invariance are related concepts is obvious enough—a smooth ball has spherical symmetry and its appearance is invariant under rotation.

As described above, in the mathematical language that particle physicists use, quantum field theory, the basic mathematical entity is the energy function (the Lagrangian or the Hamiltonian) which describes how stored energy changes when particles interact or propagate through space and time. Once we know the form of this energy, how it depends on all the components of the system, we know a lot about how the particles will behave.

Symmetries are built into quantum field theory as invariance properties of the energy function. If we construct our theories to encode various empirical facts and, in particular, the observed conservation laws, then the equations turn out to exhibit certain symmetries, or better stated, certain invariance properties. For example, if we want the theory to give the same rules of physics at all places, then the form of the energy function cannot depend on the coordinates that we use to describe position. It does depend on the values of the fields at each position, but the products of the fields that define this dependence are the same for every location. Furthermore, the form does not change when we decide to measure all our distances with respect to a different zero point.

Conversely, if we take the symmetries to be the fundamental rules that determine what theory we can write, then various observed features of particles and their interactions are a necessary consequence of the symmetry principle. In this sense, symmetries provide an explanation for these features. We will explore this point in more detail shortly.

The mystery of the missing antimatter, in addition to being one of the most intriguing puzzles of nature, provides us with an excellent lesson on the power and usefulness of the concept of symmetries. One does not need to look into the details of electromagnetic interactions, or understand the mathematical intricacies of the Dirac equation that describes them so well. It is enough to note that there are two symmetries obeyed by electromagnetic interactions (and therefore two corresponding invariances of the Dirac equation) that prevent us from solving the mystery in any other way but initial conditions, whether we like this answer or not. In other words, only if these two symmetries are violated in nature can we hope to solve the mystery in another way.

One of these symmetries is CP. With this symmetry the laws of nature would be the same for matter and antimatter. We have already told you that experiments proved that indeed CP is not an exact symmetry in nature, so we must build a theory where that is so. The other symmetry that we must break is that related to baryon number conservation. Let us first review what these words mean, in the context of our modern particle theory.

Think of the annihilation of a quark–antiquark pair. Such a process reduces the number of quarks by one. At the same time it also reduces the number of antiquarks by one. But the difference between the number of quarks and the number of antiquarks remains the same. Now think of the production of a quark–antiquark pair. Such a process increases the number of quarks by one. It also increases the number of antiquarks by one. Again, the difference between the number of quarks and the number of antiquarks remains unchanged.

The difference between the number of quarks and the number of antiquarks is called the baryon number. More precisely, the term baryon number is defined as the number of baryons (protons and neutrons and other like particles) minus the number of antibaryons (antiprotons and antineutrons, etc.). Since each baryon is a bound state of three quarks, and each antibaryon contains three antiquarks, the baryon number is really one-third of the quark number, that is, the difference between the number of quarks and the number of antiquarks. The concept of baryon number conservation predates the recognition that baryons are made of quarks, so physicists continue to use that term. So we will use the term baryon number, even when we discuss quark processes, with the understanding that in the latter case what we mean by the baryon number is precisely the quark number divided by three. Pair production and pair annihilation conserve baryon number.

Now think about the evolution of the Universe. Suppose we start from a situation where the number of quarks is precisely equal to the number of antiquarks. In such a Universe, the baryon number is zero. If all that happens to quarks and antiquarks is pair annihilation and pair production or, more generally, processes that conserve baryon number, then there can never be a time in the history of the Universe when the baryon number is different from zero. In other words, the Universe can only evolve from a state of zero baryon number to the present state, where the baryon number is nonzero (there are many baryons but very few antibaryons) if there are processes that change baryon number. (An obvious point, but instructive nonetheless.)

## Symmetries and Conservation Laws

There is a beautiful and deep connection between conservation laws and symmetries. This relationship was first understood and proven in 1915 by a mathematician, Emmy Noether (1882–1935). It is thus known as Noether's theorem. The life of Emmy Noether is a fascinating story in itself. She was the daughter of a mathematics professor at the University of Erlangen, but even so was not allowed to register as an undergraduate student there; only male students were accepted. She was permitted, however, to audit courses, and then to take the exam that marked the culmination of undergraduate training. She passed and hence was admitted as a graduate student, and five years later received her doctorate. However, the route to becoming a lecturer in a German university was not open to a woman, so she worked without pay, standing in for her ailing father to give many lectures and continuing to do research. The famous mathematician, David Hilbert, who thought very highly of her work, invited her to join his group in Göttingen. Again she began to work there with no official position and no salary. Hilbert tried to obtain a position for her in the University of Göttingen, declaring: "I do not see that the sex of a candidate is an argument against her admission as Privatdocent. After all, we are a university, not a bathing establishment." Eventually his logic was accepted and she began to be paid for the work of teaching that she had previously been doing without any salary, though her title and her pay were not those that a man with the same skills would have had.

In 1933, when the Nazis came to power, her achievements as a mathematician again had no effect when, being a Jew, she was dismissed from the University. She managed to emigrate to the United States where she became a full professor at the women's college Bryn Mawr. There her gender was viewed as an advantage rather than a drawback. (It is perhaps not accidental that this small college today typically graduates more women with degrees in physics each year than any other college or university in the country.)

Now let us examine which symmetries our theory must have. We have found many conservation laws that describe our observations. What Noether's idea tells us is that, to incorporate these laws in our theories, we must build in the related symmetries.

## Space-Time Symmetries

At the very foundations of science in general, and physics in particular, lies the assumption that the laws of physics are the same everywhere and at all times. The invariance of the equations that encode this property, invariances with respect to arbitrary choices for the origin and direction of the coordinates used to describe positions and times, must be built into the mathematics of any theory. Noether's theorem tells us that it has major implications for physics—it predicts conservation laws, conservation of energy, conservation of momentum, and conservation of angular momentum. These might have been thought to be just empirically observed properties of nature; Noether tells us that we cannot give them up without giving up the universal applicability of the laws of physics. Since that is the fundamental assumption of all of science, you can see why Pauli was so loath to abandon these conservation laws when beta decay seemed not to obey them.

Conservation laws are powerful because they restrict the possible outcomes of any process: only those that obey the law can occur. In constructing mathematical theories, symmetries have a similar power; they restrict the possible forms of the mathematical expression that describes the energy in terms of the fields. They thus greatly simplify the job of exploring the possible theories. No wonder physicists like them so much!

For space-time symmetries, it is easy to understand what we mean by invariances of the equations. Think of an American experimentalist measuring the force exerted by two electrons in his laboratory on each other. At the same time, a European experimentalist duplicates the experiment in his laboratory in Rome. The basic assumption of physics is that the two physicists should get the same result, if they properly isolate their experiment from the effects of local conditions. Now consider the mathematical equation that describes the force between the two electrons. If the two measurements are to yield the same result, the energy function cannot depend on the distance between Rome and the electrons. The only possible appearance of space coordinates in the energy function is through dependence on the relative distance between the electrons, not on their individual locations. This may sound almost trivial to you. But it is not, and it has surprisingly powerful consequences.

Invariance of the laws in time leads to conservation of energy. Invariance with respect to the origin of the coordinates describing position leads to conservation of momentum. Invariance with respect to the orientation of these coordinates (this is the symmetry under rotations to which we alluded before) leads to conservation of angular momentum. Thus we have a symmetry-related explanation of why energy, momentum, and angular momentum are observed to be conserved in all physical processes.

Conversely, the observed conservation of energy, momentum, and angular momentum encourages us to believe that the basic assumption that we make in physics is correct: results of experiments depend neither on *when* the experiment is performed, nor on *where* it is carried out. The laws of nature do not change with time or with place.

We started this section by saying we must assume these three space-time symmetries, that is, the invariance of the mathematical equations under translation in time, translation in space, and rotation, in order for there to be any point in trying to study the laws of physics. So of course we require that they are built into all quantum field theories that could describe nature. Consequently, conservation of energy, momentum, and angular momentum are features of all interactions predicted by these theories.

## Gauge Symmetries

Like the conservation of energy, momentum, and angular momentum, the conservation of electric charge is related to an invariance of mathematical equations. But this invariance is of a rather different kind. Rather than being an invariance with respect to changes of the coordinates that describe space and time, it is an invariance with respect to certain changes of the quantity that determines the electric and magnetic fields in the theory.

Remember that we told you that the photon field that appears in our energy function encodes the electric and magnetic fields, but is not itself either of them. It turns out that this encoding is not unique; there are multiple choices for how it is done, but nothing physical can depend on differences in the photon field that give the same electric and magnetic fields. Thus we have a field-parameterization invariance rather than a coordinate-parameterization invariance.

These more abstract invariances are a key feature of particle field theories. They can greatly restrict our choices in writing down the theory. Once we know that a relationship between conservation laws and such invariances exists, we can ensure that our theory has a particular empirically observed conservation law simply by building a related invariance into the equations that describe the theory. You do not need to know how to do this, but it does help to understand that it can be done and it gives us a very compact way of encoding physical properties.

But how do we know what symmetries a theory has? What defines the equations we have so glibly mentioned? We have said that the key to writing a field theory is to write an expression that describes how energy depends on the values of the various fields at each point in space. We called this quantity the energy function. The importance of this function was first understood in the nineteenth century, in classical mechanics. This understanding was extended to a field theory in classical electromagnetism, and the formalism later became the basis of quantum field theories. So the symmetries we speak of are the invariances of the energy function under certain redefinitions of the variables, the fields and their coordinates, which appear in this function.

One possible way we could change the fields without changing the energy is to make the same change at every point in space and time. Invariance under such a change is called a *global* symmetry. The second, more complicated, possibility is to find a way to make a change that is independently defined at each point in space and time, and yet does not change the energy. This is called a *local* or *gauge* symmetry. (The name has historical roots; gauge as in measure or scale, or the gauge of a railroad. The connection of this idea to this type of symmetry is, however, not only obscure, it is almost meaningless as we understand the subject today. But the old terminology stays, even without any good reason for it.) Only very special choices for the energy function will have such a symmetry.

Conservation of electric charge is related to such a symmetry. To build such a symmetry into our field theory, the energy function must be invariant under a particular related change in both the matter fields (spin-1/2, charged particle fields) and the force-carrier fields. This can only be made to work with force-carrier fields that give spin-1 particles. Conversely, and surprisingly, we find that the only sensible theories that we can write down for spin-1 particles are theories with this type of symmetry built in. If we try to do without the symmetries (by adding terms in the theory that do not respect them) and derive predictions for measurable quantities, we get nonsensical results. Therefore, local symmetries are now a part of all field theories that are related to such spin-1 force fields and thus, in particular, all the theories that describe the electromagnetic, strong, and weak interactions.

There are further striking consequences to local symmetries. One of these is that the related interactions are universal. For example, any charged particle placed in the same field feels the same force as any other like-charged particle, whether it is a proton, a positron, or a positively charged meson. The universality means that it is enough to measure the rate of a single process that is due to a particular type of interaction, and then, since the symmetry relates all such interaction processes to each other, one can predict the rate of any other process involving the same force field.

## Discrete Symmetries

The three symmetries that we tested in our car-race gedanken experiments—P, C, and their combined action CP—belong to a different set of symmetries, the discrete ones. Rather than a continuous set of variable changes (such as the possible choices for the origin of our coordinates) that do not change the energy function these symmetries just have a single change—for example, the reversal of the sign of all coordinates. Make the change twice and you are back where you started.

There are two more important symmetries in this class—T and CPT. T stands for time-reversal invariance: are the laws of microphysics (though not of probability) the same when we reverse the time clock? CPT is the combination of charge conjugation C, parity P, and time reversal T: it exchanges left with right, up with down, forward with backward, particles with antiparticles, and reverses the arrow of time.

As we already discussed, P and C are violated in a rather dramatic fashion by the weak interactions: right-handed neutrinos and left-handed antineutrinos are entirely blind to the existence of weak interactions, but left-handed neutrinos and right-handed antineutrinos are certainly not. It is simple to write field theories that violate P and C but conserve the combination CP. We can choose, for example, our theory to include matter fields that correspond to left-handed neutrinos and right-handed antineutrinos, but simply leave out right-handed neutrino and left-handed antineutrino fields. (This is precisely what we do in the Standard Model for massless neutrinos.) Such a theory cannot have either parity or charge conjugation symmetry, but it can have CP symmetry. Our Standard Model violates parity and charge conjugation, by the choice we make to leave out these additional fields, or at least to avoid involving them directly in any interactions (other than gravity when we get around to putting that into the theory).

CPT symmetry, unlike C and P, is not a matter of choice. This combination has a unique status among the discrete symmetries. All field theories have this symmetry. There is just no way to write a theory that has all other required properties, such as local interactions and coordinate invariance, but does not also have this symmetry. It has also withstood all experimental tests. Since we do not know how to write a CPT-violating quantum field theory that would make sense, any experiment that showed it to be untrue would have far-reaching consequences. It would certainly set theorists scrambling for new ideas.

The special status of CPT has implications for the discrete symmetry that is of most interest to us, namely, CP. The question of time-reversal invariance becomes intimately tied to the question of CP invariance. Break one, and in any field theory you must break the other.

Indeed, we find that it is not as easy to write down a theory that violates CP as it is to write down a C- or P-violating one. If our field theory has, for example, left-handed neutrino fields then—so requires CPT—it *must* include right-handed antineutrino fields. One cannot write a theory that violates CP by simply choosing which fields to include and which not to. It takes a detailed examination of the interactions of left-handed particles and right-handed antiparticles to tell us whether CP is respected or not. Theories with a small number of particle types are often CP conserving. The restrictions on the allowed terms from the required symmetries do not leave room for CP-breaking terms. The theory must have quite a few different particle types in it before it has a rich enough structure that CP symmetry is not automatic, just like CPT. In the time of Pauli and Dirac it would have been a huge conceptual leap to imagine there could be so many different particle types.

But there has been much progress in the theory of elementary particles since those days. Field theories and their symmetries allow us to understand many subtle features of the various interactions. We have observed many new types of particles. In describing their properties, we can write theories that, unlike the Dirac equation for electromagnetic interactions, can accommodate laws of nature that are different between matter and antimatter. Indeed experiment has shown that CP symmetry is violated, so we had better build a theory including that effect.

There are certain special cases where you can interpret the consequences of the discrete symmetries as conservation of a related quantum number.

These are cases where the system under study both starts and ends in a state that is unchanged by operation of the symmetry. But not all states are of this type—any charged particle transforms into a different particle under C for example.

So, in general, discrete symmetries do not imply any conservation law. Still, they tell us important things about what might happen and what might not happen in various processes. For example, when we discussed cobalt decays, we said that parity would imply that the number of electrons emitted in a direction parallel to the spin of the $^{60}$Co nucleus should be equal to the number of those in the direction opposite to the spin. There is no conserved quantity in that statement, but there is still an interesting restriction on the possible outcomes of experiments that is implied by the symmetry.

## Baryon and Lepton Number Conservation?

All the processes that have been measured in our laboratories seem to obey two additional conservation laws. Experimentalists have never observed a process in which baryon number or lepton number has changed, even though they have tried very hard to do so. Proton decay into a positron and a neutral pion would be allowed by all the space-time and local symmetries that we know to exist. Proton decay has never been observed and it is not for want of looking for it! Indeed, experiments have set an incredibly strong limit on its rate, telling us that the half-life for proton decay is greater than $10^{33}$ years. Other processes that would require a change in baryon number, such as the chance that a neutron becomes an antineutron, have also been searched for and not found, though the limits are a little less stringent.

Is this conservation the result of an additional symmetry? Or is it just an accidental effect of the symmetries we have already imposed? Amazingly enough, this last possibility is what happens in the Standard Model. Given its known symmetries and the list of particles, baryon number conservation arises as an accidental symmetry. Actually, what happens in the Standard Model is even more exciting than that. For the electromagnetic and strong interactions, the baryon and lepton number conservation law is exact. But for the weak interactions, the symmetries related to baryon and lepton number are not exactly respected. They are broken but only by very special terms in the energy function. These terms are such that for any process that happens at low temperature, the violation of the symmetry is very, very (very) tiny. It is so small that in everyday language one would say it simply doesn't happen.

The tiny processes that break these conservation laws are so slow, so the theory says, that we can never observe them in experiments on matter at ordinary temperatures and densities. But if we were able to carry out experiments at higher temperature and density, then, so the theory tells us, the breaking of this apparent conservation law would be much more prominent. Processes that can change baryon number and lepton number (though not their difference) would occur much more readily under such conditions.

# STANDARD MODEL
# GAUGE SYMMETRIES

So now we see that the basic building blocks of the Standard Model are its symmetries. Remember that when physicists say symmetries they mean the invariances of the energy function under certain redefinitions of the fields or their coordinates. These symmetries were first noticed through empirical discovery of properties such as conservation laws.

## The Symmetry behind the Electromagnetic Interaction

Let us begin again with the familiar case of electromagnetism. Here there are three crucial observational facts: Any electric charge (of a particle in isolation) could be quantified in integer (or whole number) multiples of the charge of one electron. There is a universal force law between any two electric charges (the force between them scales as the product of the two charges divided by the square of the distance between them). Electric charge is conserved; it is neither created nor destroyed in any observed process.

These basic properties, together with further observational facts, such as the relationship between electric and magnetic effects, led the physicists of the nineteenth century to a field theory that described both electricity and magnetism in a unified way. They could find an energy function based on an abstract, space- and time-dependent field, which we now think of as the photon field. The formalism gave Maxwell's equations as its predictions for how charges and electric and magnetic fields are related, and how they move or change. The Dirac equation, combined with this field theory, led to a quantum field theory of electromagnetism, known as quantum electrodynamics (QED). This is the archetype of all gauge quantum field theories.

The amazing thing is that, having determined these properties, physicists could indeed find an energy function incorporating a symmetry or invariance that would "explain" them. The mathematics seems at first a little arcane. Remember that gauge invariance is an invariance of the energy function under correlated redefinitions of both the charged matter fields and the quantity that encodes the electric and magnetic fields, which we call the photon field. A remarkable special feature of gauge invariance is that these redefinitions can be made differently for each point in space and time. We call this a local invariance. The form of the energy function is unchanged when the fields are redefined in the appropriate way. Not only that, but also the values of the electric and magnetic field derived from the photon field are unchanged when it is redefined in this particular way.

As you might guess, the requirement of gauge invariance is a very powerful restriction on the form of the energy function. It thus has powerful physical consequences. In particular, it requires that the force-carrying particles associated with these interactions are massless. There is no way to write a term in the energy function that would represent a mass for the photon that is invariant under these field redefinitions. The gauge invariance in fact requires that the photon is massless, carries spin-1, and has no electric charge. (This, in turn, predicts the Coulomb force law: as the distance $r$ between two charged particles changes, the force between them changes as $1/r^2$.)

Notice that a universal coupling strength does not mean that all particles have exactly the same electric charge. The symmetry can accommodate any definite charge assignment for all particles in a given class. The charge of a proton or an electron is the smallest charge observed in isolation, and, by convention, we still call that one unit of charge, +1 for the proton and −1 for the electron. Once quarks were understood we recognized that down-type quarks have the smallest charge, −1/3 in proton units. The reason we use the proton and not the quarks to define the basic charge unit is that we only knew about quarks long after we had adopted the convention to call the charge of the proton +1 and that of the electron −1. The fact that quarks are confined, and hence never observed in isolation, fooled us into thinking that this was the smallest unit of charge. The fact that quarks have fractional charges in these units was one of the biggest barriers to acceptance of the quark idea when it was first suggested, since experiments had never observed anything with a charge smaller in magnitude than 1 in proton units.

The form of the invariant terms also restricts the form of electromagnetic processes. Any particle with charge can emit or absorb photons, but does not change type when it does so. The only way new particles can appear is in pairs, with matched particle and antiparticle. Likewise, only matching particles and antiparticles could annihilate, disappearing to leave only some photons.

Thus, if electromagnetism were the full story, we would have not just conservation of electric charge, but a separate conservation law for every individual particle type (number of particles minus number of antiparticles). Moreover, the laws of physics would be identical for matter and for antimatter. All of these results follow from the restrictions imposed by the gauge symmetry. It is remarkably powerful. Indeed, you can already see that it cannot be the full story, because it is too restrictive.

## The Symmetry behind the Strong Interaction

In the late 1960s to early 1970s it was recognized that strong interactions, like electromagnetism, could be understood as a theory of spin-1 particles. Let us review that history a little. Quarks were first postulated as constituents that could explain the patterns of masses and charges of hadrons. That original idea and its observational confirmation in high-energy electron-scattering experiments left open the question of the forces between quarks. Once again a set of observational facts begged for a symmetry explanation. Why do quarks combine in sets of three to make baryons, or quark plus antiquark to make mesons? Why are there no four-quark or two-quark hadrons? Indeed, the biggest mystery for the quark idea was that quarks, and their fractional electric charges, are never found in isolation.

The answers lie in the symmetry properties of the theory that describes the forces between and among quarks. This theory introduces a more complicated set of field redefinitions than that of QED. The charges that define how quarks fields transform under this redefinition and how quarks couple to strong force fields are called color charges. Because of the role of color charge together with the similarity of the mathematics to that of quantum electrodynamics, physicists coined the term quantum chromo-dynamics (QCD) to name this theory.

Gluons are the force-carrying particles of QCD, the strong interactions of quarks in the Standard Model. The invariances of the theory require that gluons, like photons, are massless, spin-1 bosons. However, because of the more complex algebra of invariances in this theory, gluons also carry color charges. This is unlike the situation for electromagnetism, in which the photon has no electric charge. This means gluons interact strongly with each other, as well as with the quarks, and that they too are confined particles.

One very important requirement that follows from the gauge invariance is that the strong interactions must be universal. By *universality* we mean that the strength of the strong interaction between quarks, mediated by gluons, is the same for all quark flavors. In terms of the energy function, the gauge invariance requires that, independent of which quark field is involved, all terms that couple quark fields to gluon fields must have the same numerical coefficient, which we call the coupling constant. The invariance requires this universality in the coupling. The size of this coupling sets the scale of the energy due to these terms. Hence it sets the size of strong forces and defines how rapidly any strong interaction processes occur. It gives the Standard Model a very strong predictive power. If you measure the rate of a single strong interaction process, you can fix the value of the strong interaction coupling between quarks and gluons, and then predict the rate of all other gluon-mediated processes. (In practice this is applicable only for high-energy processes; the calculation just gets too complicated at lower energies.) This predictive power applies also to only-gluon processes (with no quarks), once the different type of color charge that gluons carry compared to quarks is taken into account.

Gauge invariance requires universality not only in the strong interactions, but also in the electromagnetic ones. Here, however, the predictive power is weakened because we can a priori assign a different electromagnetic charge to each particle species (this is impossible for color charges).

But once we assign the same charge to, for example, electrons and muons, gauge invariance requires that the strength of the electromagnetic interaction between an electron-positron pair, mediated by the photon, is precisely the same as between a muon–antimuon pair. It is easy to test the predicted universality of interactions that are related to gauge invariances, that is, all interactions mediated by spin-1 force carriers. It has been beautifully confirmed in numerous experiments.

The symmetry of QCD (mathematically, it is called SU(3)) requires that each flavor of quark can have any of three different possible color charges. (Remember, flavor is the generic word for the names that distinguish quarks with different masses.) Likewise, antiquarks come in three anticolors. Gluons, it turns out, must carry both a color and an anticolor.

Neither quarks nor gluons are found in isolation; they are found only in color-neutral combinations, which are the things we call hadrons. Color neutrality explains the basic quark structure of baryons (three quarks, one of each color), antibaryons (three antiquarks, one of each anticolor), and mesons (one quark and one antiquark, which must have matching color and anticolor and equal probability of any one of the three colors); these are the three possible ways to make a color-neutral object from quarks and antiquarks.

Like photons from electrically charged particles, gluons can be emitted and absorbed from any particle that has a color charge. While emitting a gluon can change the color-charge state of the emitting particle, it does not change its flavor type. Similarly, quark–antiquark annihilation into gluons or production from gluons occurs only for matched particle and antiparticle flavors. Color quantum numbers too must be conserved, but they can be passed from quarks to gluons and back. For example, a red quark can become a green quark by emitting a red-antigreen gluon.

Once again the gauge symmetry restricts us to a world with an exact symmetry between matter and antimatter, exactly the same strong interaction physics for quarks and antiquarks, and with separate particle minus antiparticle number conservation laws for each flavor of quark. Further, again like QED, we have constructed a theory in which mirror symmetry or parity is exact. All these properties fit our observations about strong interactions. (There is a technical exception to the automatic P and CP symmetry of this theory, but observations tell us that this effect is so small that we can safely ignore it here.)

We can say that the symmetry principle explains all the empirical facts. What we mean is that all these properties follow when we write an energy function based on space- and time-dependent fields which transform in defined ways, collectively known as the algebra of SU(3), and insist that overall the energy of any system must be invariant with respect to this set of transformations. Of course we picked SU(3), and not some other symmetry group, because we noticed that baryons are made from three quarks and then looked for a theory that fit.

## The Symmetry behind the Weak Interaction

The third type of interaction encoded into the Standard Model is the weak interaction. It is so weak that forces due to it can only be observed in very special situations, where only these forces can play a role. Weak interactions are important, and were noticed, because they allow processes to occur that would otherwise be forbidden. They have a different symmetry structure and so there are conservation laws that are respected by the strong and electromagnetic interactions but not by weak interactions.

The first example of a weak interaction is beta decay, the process in which a neutron turns into a proton plus an electron and an electron-type antineutrino. This is the archetype of charge-changing weak transitions. (The overall charge is, of course, conserved. The charge change occurs between the two quarks, and is compensated by an opposite charge change between the two leptons.) At the quark level one can describe this process as a down quark becoming an up quark by emitting the carrier particle of the weak interaction—a W-boson—and then the W-boson very rapidly decaying to produce the electron and the electron-type antineutrino.

In strong and electromagnetic interactions quarks emit and absorb gluons or photons but they never change flavor type in so doing. But the *W* carries electric charge. So it is clear that *W* emission or absorption *must* change flavor type if it is to conserve electric charge. The *W*-mediated or charge-changing weak interaction is the mechanism that makes all the more massive quark flavors unstable. This in turn makes all the hadrons built from more massive quark flavors unstable, leaving the proton as the only stable hadron (though neutrons do not decay when they are in a stable nucleus). This is why particles containing second- and third-generation quark flavors, which are produced in high-energy accelerator experiments and were common in the hot early Universe, are not found in ordinary matter. Likewise, the charged leptons of the second and third generation, the muon and the tau, decay via weak interactions.

There are many possible flavor-changing weak interactions. Any up-type quark (i.e., any quark with charge +2/3) can emit a W-plus boson or absorb a W-minus boson. By doing either, it changes into a down-type quark (i.e., a quark with charge −1/3). Any down-type quark can emit a W-minus boson or absorb a W-plus boson. By doing either, it changes to an up-type quark.

This pattern is more complicated compared to all other gauge interactions, and that carries over to particle–antiparticle pair production or annihilation. Again, because of its electric charge, when a $W^\pm$ produces a particle and an antiparticle, they cannot have opposite electric charges, because the charge of the $W^\pm$ cannot just disappear; so the particle and antiparticle must be different flavors. Any pair that satisfies charge conservation (and of course energy and momentum conservation too) can be produced.

At first glance there appears to be no universality in the rates of quark weak transitions; different flavor changes proceed at different rates. For example, strange quarks undergo beta decay much less rapidly than down quarks. Since we stressed the universal pattern of gauge interactions, you may well wonder how we can we fit these observations into a consistent gauge-invariant theory. Indeed, the universality does hold also in weak interactions, but it appears there in a more subtle way. We will explain it later; at the moment it tells us that we still miss a piece in this puzzle.

Leptons too can emit and absorb W-bosons. If, for the moment, we ignore neutrino masses, we find a universal pattern for lepton transitions. The tau decay into a muon and a neutrino–antineutrino pair depends on precisely the same coupling constant as the muon decay into an electron and a neutrino–antineutrino pair (and it is the same weak interaction coupling constant as that of the quarks). But if neutrino masses are included, the lepton patterns are seen to be as complicated as those for the quarks. This fact gives you a hint that the universal pattern of weak interactions is realized in nature, but it is somehow hidden or disrupted by the masses of the quarks and leptons.

The first big success of the modern gauge theory of weak interactions was that it predicted the existence of Z-bosons and of the weak interactions mediated by them. Indeed, the gauge invariance that is related to the weak interaction predicts that there are three, not two, spin-1 force carriers. Like strong and electromagnetic interactions, the $Z^0$-mediated processes do not change flavor. (Indeed, it was this feature that first suggested a fourth quark; with only three quarks there was no way to avoid $Z^0$ couplings that changed flavor, $s$ to $d$, and this was ruled out by experiment.) But, like $W^{\pm}$-mediated processes, they are different for left-handed and right-handed states. Also like the $W^{\pm}$, the $Z^0$ bosons are massive spin-1 particles, not massless like the photons and gluons. $Z^0$ couplings are different for up-type and down-type quarks, charged leptons, and neutrinos, but they are universal in the sense that, like the couplings of the photon, they are generation independent. Furthermore, like the photon and the gluons, the Z-boson couples only to matched particle–antiparticle pairs.

Soon after the theory predicted $Z^0$-mediated processes, they were seen to be present in experiments. They had not been seen before mostly because no one had really looked for them. Physicists knew that there were no weak interactions that changed flavor without changing charge. The idea of weak interactions was originally so bound up with the phenomenon of flavor change that no one ever stopped to ask whether there could be processes that change neither flavor nor charge, but do not respect parity symmetry. But, once one tried to write a gauge theory for weak processes, such an effect could not be avoided. Thus observation of weak processes mediated by the neutral Z-boson was the first major confirmation of the gauge theory of weak interactions.

# 14

## A MISSING PIECE

### The Puzzle of Particle Masses

There is a big puzzle in the Standard Model. The W- and Z-bosons have spin-1. Spin-1 force-mediating bosons require that we build a gauge theory—a theory with a gauge (or local) symmetry, which we described in chapter 12. We know no other way to get a well-behaved theory for spin-1 particles. But we also know that gauge theories predict massless spin-1 bosons and universal coupling. The $W^{\pm}$ and $Z^{0}$ are massive and the W-boson does not seem to have universal couplings with the various quark flavors.

Furthermore, in order to reproduce the observed parity violations of weak interactions, the Standard Model assigns different weak charges for the left-handed and right-handed matter particles. This has a price. The price is that the Standard Model as described above predicts that the quarks and leptons are all massless. The only way to write terms in the energy function that cause masses for spin-1/2 particles is to mix together their left-handed and right-handed parts. A single massive spin-1/2 particle has both a left-handed and a right-handed component. But if the left- and right-handed fields have different symmetry properties then we cannot write a mass term that does not violate the symmetry.

So a mass term is inconsistent with a gauge interaction that just affects left-handed states. It breaks the invariance of the equations. Either we give the same interactions to the left-handed and right-handed components

140

of the particles in question, or we leave out the mass terms, that is, set them to zero.

Experiments tell us that parity is violated in the weak interactions, so left-handed and right-handed particles indeed must have different weak interactions. But experiments also tell us that all quarks and leptons are massive. We seem to have some observations that cry out for a gauge theory and others that are at odds with what such a theory would predict! The answer to this impasse is quite a surprise. It turns out we can write a theory that has all these symmetry properties and then contrive to find a way to keep all the good predictions and remove all the wrong ones. The trick is spontaneous symmetry breaking.

The symmetry exists in the equations that describe the physics, but nature, in finding the lowest-energy state, must make a choice, like the donkey between two buckets, that breaks this symmetry. A fifth and quite different type of interaction force field, together with its own new particle type, must be added in the Standard Model to achieve this effect. This added force field is called the *Higgs* field. It is responsible for many of the observed features of weak interactions, including the CP-violating K-meson decays.

The name Higgs comes from Peter Higgs, a British physicist, who was one of those who contributed to explaining how spontaneous symmetry breaking could work in gauge theories. He was not the only one; pieces of this idea were already understood. But Higgs added a particular insight to solving this puzzle, putting together the disparate pieces that others had suggested to show how a theory could have all the desired properties. Physicists soon took to using his name to describe the fields that drive the spontaneous symmetry breaking. (It helped that Higgs had a simple name, the Anderson-Kibble-Brout-Englert-Higgs effect is just too much of a mouthful, although all these others contributed important pieces of the idea that goes by the name of Higgs.) Higgs is a modest man, quick to point out that many others contributed the basic ideas. He speaks of "the so-called Higgs field." However, sometimes it is not something new that is needed, simply a new insight on how the pieces can be put together in a new way, and that is what Peter Higgs gave us.

We will soon learn more about the important role that the Higgs theory plays in the story of CP violation, but let us first try to explain the role it plays in giving mass-energy to all the particles that we observe to have mass. To do this we have to introduce you to the notion of nothing, as it appears in field theory. The lowest-energy state in a field theory represents the state with nothing in it, the physical vacuum. If the lowest-energy state in the theory is a state that does not have the full set of symmetries that was present in the equations, then some symmetry is spontaneously broken.

The idea that the lowest-energy state of a system may have less symmetry than the equations that describe the system may seem a bit far fetched. But actually it is true in many familiar situations. Imagine a round broomstick balanced vertically. The equations that describe this system have complete symmetry with respect to rotations about the center of the broomstick. But the lowest-energy state of the system has the broomstick lying on its side on the ground. In falling over it must fall in some direction. Even in a vacuum, where no wind blows, quantum theory tells us that it must eventually fall. (Everything that can happen must eventually happen.) The probability it will fall is equal for all directions (figure 14.1). When it falls, it spontaneously breaks the symmetry that existed when it was standing upright.

In the case of the weak interaction field theory, we must add something to the theory to achieve this trick—some broomstick or donkey that will choose a way to fall and thereby break the symmetry. The simplest way to do that is to add a single new type of field, representing a spin-0 (spinless) particle. This field is called the Higgs field. In order to affect the weak interaction symmetry at all, this field must couple to the W- and Z-bosons. In order to fix the problem of quark and lepton masses, it must also couple to quarks and leptons.

The observations tell us that we definitely need some such spontaneous symmetry breaking to occur. But so far there is no direct observational confirmation that adding a Higgs field is the right way to achieve it. Unlike the rest of the Standard Model, this feature still has somewhat provisional status. Any other way to allow the spontaneous symmetry breaking is more complex, so, unless some experiment rules it out, we will assume that a single Higgs field is the answer. We devote the rest of this chapter to explaining its role in the theory. Certain aspects of this role, namely, the way that CP violation appears in the theory via the Higgs couplings to fermions, are critical for the history of matter and antimatter in the Universe.

Fig. 14.1 The probability that the broom-stick will fall is equal for all directions.

Once we add a Higgs field to the theory, we must examine what kinds of interaction terms that field can have, consistent with all the symmetries. The Higgs interactions in the Standard Model are very different from the gauge interactions. The field associated with a spinless particle, unlike electric and magnetic fields, has no direction to it. A Higgs field simply has a magnitude at each point in space. This allows a greater variety to the terms in the energy function that involve this field.

There must also be terms in the energy function which define the energy due to the Higgs field itself. Because these are spin-0 particles, they have different rules from the spin-1/2 and spin-1 particles we have described above. In particular, they can have no handedness even for their weak couplings, because they have no spin. So, in contrast to the case for quarks and leptons, the symmetries allow a term in the energy function which makes the Higgs boson massive. There is nothing in the theory to tell us what its mass should be.

At present, all we can say is that the experiments would have seen effects from a single Standard Model Higgs boson if it existed with a mass of less than about 120 times that of the proton, so we know its mass must be heavier than that. However, some very precisely measured quantities are sensitive to indirect effects of the Higgs particle. We can calculate these effects, and our result depends on the Higgs mass. Unless this mass is less than about 200 times the proton mass, our calculations do not match experimental results. We need some indirect Higgs-like effects to make all the observations fit together. So either experiments will find the Higgs particle soon or we will have to modify the theory in some way.

The interaction of the Higgs field with quarks and leptons is of a type known as a *Yukawa interaction*. You may recall that we said that Hideki Yukawa predicted spin-0 particles, the pions, emitted and absorbed by spin-1/2 particles, protons and neutrons, as an explanation of strong interactions. His name is now used for any process involving the emission or absorption of a spin-0 particle by any spin-1/2 particles. In the Standard Model, the Yukawa-type interactions are couplings between the Higgs field and the quarks or leptons. Any generation of quarks can combine with any generation of antiquarks to couple to the Higgs. Likewise there are Yukawa-type terms that couple the Higgs field to the leptons.

We have stressed that the symmetries of the gauge interactions, those of the spin-1 force carriers or gauge particles, impose the requirement that the couplings of the gauge particles to all fermions have a single universal strength parameter. Now, while gauge interactions must be universal, the Yukawa interaction has no such restriction from the symmetries. For each quark–antiquark (or lepton–antilepton) combination that couples to the Higgs boson, the strength of the coupling is an independent parameter. There is not even any requirement that the quark and antiquark are matched in flavor.

This means that, once Higgs interactions are added, the energy function of the theory contains a lot more terms with independent coupling constant parameters. Of the twenty or so parameters of the Standard Model (or thirty odd if we include the effects of neutrino masses), only three would remain, the three gauge couplings (one each for strong, electromagnetic, and weak interactions), if the Higgs boson and its interactions were not part of the model.

Given that the Yukawa interactions of the Higgs boson bring in so many additional parameters, and that the Higgs boson has not been observed yet, why would we want to complicate the Standard Model in this way? We must, because the Standard Model with quarks and leptons and gauge interactions alone is inconsistent with observations. In particular, as we noted earlier, the weak interaction symmetry structure of the equations would require that all the particles of the original Standard Model are massless. But in reality, all quarks and leptons have masses, not to mention the very massive W- and Z-boson masses.

So we need to add a spin-0 Higgs field (or more) to our theory to allow all these other particles to acquire masses via a spontaneous symmetry-breaking effect. This then predicts another particle, associated with this new field—which we call the Higgs particle. If there turns out to be no fundamental Higgs particle and associated field to achieve this effect, there must be some more complicated version of the idea in play, with some kind of spin-0 particle, perhaps a composite one, or perhaps more than one. Something of this type is essential to provide the spontaneous symmetry breaking, to make the donkey's choice, and thereby allow masses for the fermions into the theory without totally removing the gauge symmetry that so effectively gives the spin-1 bosons.

We have not yet seen direct evidence for Higgs particles, but we are quite convinced that they, or something like them, must exist. Can these neutral and weakly interacting Higgs particles be the dark matter? The answer is *no*, Higgs particles cannot constitute the dark matter because they decay, with a half-life much much shorter than the age of the Universe, so any such particles produced in the early Universe would no longer be around to form the galactic halos of dark matter.

All quarks and charged leptons get mass because they feel the effects of the Higgs field due to their Yukawa couplings with it. But how much Higgs field do they feel? The mass of a particle is just the energy associated with an isolated particle—a particle at rest, alone in the vacuum. So the field they feel must be the vacuum value of the Higgs field. This seems contradictory—surely the vacuum is empty?—yet here we say that the Higgs field in the vacuum must not be zero in order that particles have mass. This needs some explanation.

## How Do We Describe Nothing?

In field theory, the vacuum is simply the lowest-energy state. This is what one would usually call *empty space*. This state is certainly empty in the usual meaning of the word; by definition it contains no particles. Indeed, in any sensible theory it has no structure, nothing differentiates one place, or direction, or time, from any other. But that does not necessarily mean that it is the state where all fields are zero everywhere. If we can choose the energy function so that the lowest-energy state is one where the Higgs field is not zero, then the vacuum will be a state where that is the case.

Physics only makes sense if we choose as empty space, the vacuum, the state of the system with the lowest possible energy. How could anything have less energy than nothing? So, if the lowest-energy state contains a constant Higgs field, then the vacuum state contains a constant Higgs field. Indeed, we find this is just what we need to fix the wrong predictions of massless particles and universal weak interactions.

Now this is a remarkable claim, that empty space is not empty in some strange way. It contains a nonzero but constant Higgs field everywhere. Why would a field have a nonzero value everywhere? It is simply a matter of how energy depends on this field. Suppose the energy decreases as the value of the Higgs field increases, up to some value we will call $v$, and then increases again if the field value gets bigger than $v$. Such an energy function can be constructed for the Higgs field without destroying any of the desired symmetry properties of the theory. The lowest-energy solution for such a theory is one where the field value when no particles are present is $v$, rather than zero, and where particles or waves in this field are fluctuations of the field value away from $v$, instead of away from zero.

Two ants and a sombrero will help us understand how a symmetry can be hidden (figure 14.2). Imagine an ant standing just at the top of the sombrero and looking around. It can easily recognize the symmetry

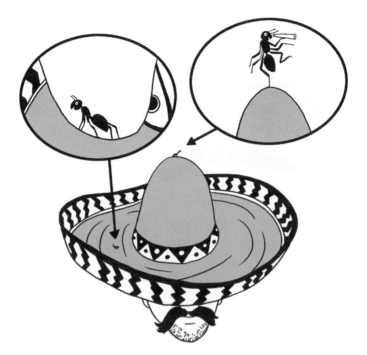

Fig. 14.2 Two ants and a sombrero.

of its situation to rotations, because the sombrero looks exactly the same in every direction. But another ant, standing at the "valley" at the bottom of the sombrero, will find it very hard to tell that there is such symmetry. To go to the top, it has to climb, but to walk sideways, the path is flat. So there is certainly no symmetry to rotations where this second ant stands. The physicist that thinks about the state at which all fields are zero is the ant at the top of the sombrero, which can identify the symmetries with ease. If, however, the ant loses its grip (in principle, quantum effects can do that), it will slide all the way down to the valley. Standing at the top is not a stable situation. The physicist thinks of "nothing" as the state of lowest energy at the bottom of the sombrero. There indeed the ant has the lowest gravitational energy—if is loses grip it will still stay in the valley. The symmetry, however, is not manifest. An intelligent ant, if there were one, could find clues for the symmetry by noticing, for example, the flatness of the valley, but it would then have to make further investigations to confirm its suspicion. Physicists deduced the existence of the weak interaction gauge symmetry in spite of its being hidden, and then investigation showed that this guess explained features of nature.

The vacuum could be any point around the valley, but whichever one we choose the symmetries are hidden and the Higgs field has the same size, because the valley is a circle at a fixed distance from the center of the hat. Because of the symmetry it does not matter which place in the valley we choose as our vacuum; the symmetry guarantees that all physical quantities are insensitive to that. Whichever way the ant fell, it makes no difference to the story.

A field related to a spin-0 particle can have a nonzero value at every point in space and not destroy any of the important invariances of the mathematics with respect to coordinate orientation. Of course, the nonzero value has to be the same everywhere in space, and at all times, or it would destroy invariance of the laws of physics with respect to location and time. Note also that directions in space have nothing to do with the direction we went in the plane where the hat lay to get to our chosen vacuum—that was a plane in Higgs field values, not in space coordinate values.

So empty space in such a theory is one where the Higgs field has the value $v$ everywhere. This constant Higgs field looks like (and indeed is) nothing to any observer. Any region with any other value for the Higgs field, including a region where it is zero, looks like something, because it contains energy density, in the form of Higgs particles. Particles exist wherever the fields differ from their constant lowest-energy values.

For any fields related to particles with spin, the vacuum must be the state where the value of all physical fields is zero. Since these fields have nontrivial properties under rotations, any other choice would give us a horrible result: empty space would have preferred directions associated with it. So, though we might contrive to write theories that have this property, observation tells us we had better choose an energy function that minimizes at zero value for these fields. So we do.

The everywhere-constant Higgs field in the vacuum is not observable as an entity of any kind. However, we do see important effects that occur because it is present. Because of their Yukawa couplings, quarks and charged leptons feel the effects of this nonzero Higgs field. They have additional energy due to its presence, energy that is there even when they are at rest. Rest energy, by Einstein's mass-energy equivalence, is nothing other than mass. Quarks and leptons gain masses from their interaction with the nonzero Higgs field.

The more strongly a particle interacts with the Higgs field, that is, the larger its Yukawa coupling, the larger the mass it acquires. Since there is nothing in the theory that fixes the size of the Yukawa couplings, and thus no reason that they should be universal, the masses of the various quarks and leptons can be different from each other, as indeed they are.

Because we added a Higgs field to the theory we also added another particle, the one we call the Higgs boson—because every type of field in a field theory must have a particle associated with its deviation from the vacuum value. (There is a very technical detail here, which in fact is what Peter Higgs told us—there is no particle at all associated with changing the Higgs field in the direction around the valley, because of the way this symmetry was linked to the gauge symmetry of the W-boson, and gave mass to that particle. The Higgs boson corresponds to changes in the radial size of the Higgs field away from the value $v$, and the mass of the Higgs boson corresponds to the curvature of the sombrero's brim at this lowest point.)

Notice that this theory presents, for the first time, an explanation for why particles have mass. In the Standard Model, mass is simply the energy due to interaction of particles with the vacuum Higgs field. No longer is it mysterious that mass is a form of energy. Like any other form of energy, the mass is determined by writing the way the fields combine in the energy function of the theory and in particular by the Higgs field Yukawa coupling terms in that energy function. The theory then tells what energy is associated with a particle at rest, and this is the mass energy (or rest energy) for that type of particle.

So far we discussed the interactions of quarks and leptons with the Higgs field in the vacuum. Would the force carriers feel the effect of this Higgs field? Only if they interact with it. Here one can turn the question around: do Higgs particles feel the effects of the gauge fields? The answer depends on the charges that are carried by the vacuum part of the Higgs field. If, for example, it carries neither color nor electric charge, then it will feel the effects of neither the color field nor the electromagnetic field. This will also mean that neither the gluons nor the photon feel the effects of the Higgs field. Consequently, they should be massless, as indeed they are. We had better arrange our theory so that whatever Higgs field component is not zero in the vacuum is indeed uncharged under electromagnetic and strong interactions; that is, not altered by the field transformations that express the symmetries of those interactions. We can do that.

On the other hand, the Higgs field has nontrivial transformation properties under the field redefinitions related to the weak interactions, so the weak gauge bosons, that is, the $W^{\pm}$ and the $Z^0$, feel the effect of the Higgs field and gain masses. Indeed, we observe that the weak gauge bosons are massive. The fact that the Z- and W-bosons are about a hundred times heavier than the proton is the reason that weak interactions have a very short range.

The Higgs field is the donkey with his head in the center surrounded by a ring of water all equally far away—or the ant on top of the hat (figure 14.3). Once the donkey makes a choice, once the Higgs field takes a particular value everywhere in the vacuum state, then we are no longer free to redefine that field by any transformations. We said before that the energy function did not change under these transformations. What this tells us is that there are in fact many different choices for the exact form of the vacuum Higgs field, all equally low energy. Any one of them will be a perfectly acceptable vacuum. But once it is chosen, the weak interaction gauge transformations can no longer be applied—their symmetry has been spontaneously broken, or hidden, by the choice of the vacuum Higgs field. Amazingly, the symmetry is broken in such a way that all the good predictions of that theory are kept—in particular, we still have a sensible theory for spin-1 force carriers—but the bad ones, such as the zero-mass particles, are removed.

So there are two ways to think about what we mean by nothing and the Higgs field. Each of them is convenient for different purposes. In the first way, we could be tempted to call nothing that state in which all fields are zero. In that description we see all the gauge symmetry and universality of the weak interactions explicitly. But when we solve the equations that describe the various fields and their interactions, we find that what we thought of as nothing is not the state of lowest energy. So it cannot be a stable situation, and hence cannot describe the ground state or vacuum state of the system. It is not nothing—even though it

Fig. 14.3 The Higgs field is the donkey with his head in the center surrounded by a ring of water.

is a state with zero value for all fields—it contains a large collection of Higgs particles.

In the second way, we call nothing the *vacuum*, or the state of lowest energy. This makes more sense because this is a stable state; it can last forever and has no structure. Physically we observe it as empty space. But mathematically this vacuum state does not look as if it is empty. It contains a nonzero Higgs field everywhere. Particles are fluctuations of the fields with respect to this state. Compared to this state, quarks, charged leptons, weak gauge bosons, and the Higgs bosons themselves are all massive. The masses arise because of their interactions with the everywhere-constant vacuum Higgs field. This language is more physical. It is convenient for a discussion of experimental results, since we usually identify particles in an experiment by their masses.

On the other hand, some of the symmetries that are the very basis of the Standard Model are not at all easy to identify in this language. In particular, the weak interaction gauge symmetry is not manifest. We will continue our discussion in this more physical language, in which the nonzero Higgs field is incorporated into the definition of the vacuum state, and the interactions of particles with this everywhere-present field are incorporated into the definitions of particle masses and couplings. In this framework, the quarks and the leptons, as well as the weak interaction W- and Z-bosons, are massive particles.

The hidden symmetry idea explains why couplings of the W-bosons to quarks do not have manifest universality. As long as we do not consider the Yukawa interactions with the Higgs field, there is no way to distinguish the flavors; indeed it would take considerable cleverness to find out there are three generations. Let us add the Higgs couplings step by step and see how the game works. First we will add Higgs couplings only for the up-type quarks. Then we will see that there are three of them, and that each has weak decays that produce a distinct massless down-type quark—just as we see that the three charged leptons have weak interactions that produce a distinct neutrino type. Let us name these the three down-flavor states. We will also still see universal weak interaction strengths, the same for all three up-type quarks, just as we do for all three charged leptons.

Now let us add Higgs interactions for the down-type quarks. We can do this in a way that mixes up the three down-flavor states, so that each down-type quark of definite mass is a particular mixture of the three flavor states, and thus is coupled by $W^-$ emission (or $W^+$ absorption) to all three up-type quarks. But now if we look at the weak interaction of a single up-type quark and a single down-type quark of definite mass, we see no universality, because it depends on how much of that particular flavor is in the down mass state. Only when we include the coupling of a single up-type quark to all down-type quarks does the universality reappear.

Think of a cake, divided between three friends in precisely equal portions. Thus, the division of the cake between Peter, Paul, and Mary is universal. But if the cake is made of vanilla, chocolate, and strawberry flavors, Peter may get a lot of vanilla, a little chocolate, and even less strawberry, while Paul and Mary enjoy more of the latter two. If we ask how much of each flavor did Peter, Paul, and Mary get, we see no universality. But if, for each of the three friends, we add up the three flavors, we soon realize that justice was maintained. The universality is hidden, but it did not disappear.

In analogy, the knife that divides the cake into three is doing so with the universal $W^{\pm}$-mediated weak interaction. But the cook that spreads around the three flavors is doing that with the nonuniversal Higgs-mediated Yukawa interaction. When experiments measure weak interaction rates for particles of well-defined masses, there is no apparent universality. But if you sum over all final particles, you will uncover it again. For example, the rate of muon decay to a muon-neutrino and a $W^{-}$ (which then produces, say, an electron and an anti-electron–neutrino) has a very different coupling from that of top-quark decay into a strange quark and a $W^{+}$. But if one examines the total rate of top-quark decay into any down-type quark plus the W-plus boson, one finds universality: it is one and the same coupling that determines both rates.

## At Last, CP Violated in the Standard Model

We have described how local (gauge) symmetries are the basic building blocks of the Standard Model, and the features of these theories that predict and constrain the patterns of strong, electromagnetic, and weak interactions. Parity P and charge conjugation C are violated in weak interactions, but conserved in the strong and electromagnetic ones. The looking-glass world is not quite the same as the one outside the glass. So of course the theorists built a theory that followed this pattern. Even then the theory had automatic CP symmetry. A theory with only gauge interactions, that is, interactions mediated by spin-1 particles, such as the Standard Model before adding in the Higgs, always respects CP. This is a consequence of the fact that the spin-1 particles couple always to matched particle-antiparticle pairs and, furthermore, that their couplings are universal.

But then we added the Higgs field and its many different Yukawa couplings, and chose the energy function so that this field must have nonzero value in the vacuum. What happened to CP symmetry in that step? Did we fix the CP problem as well as the problem of particle masses? Did we achieve a theory that could accommodate the observation that CP symmetry is not exact, that the laws of physics are not identical for matter and for antimatter? The answer to that question required some further detailed considerations about the Yukawa couplings of the Higgs field and their consequences.

When the Standard Model was first formulated, it was a two-generation theory, with just four quark types and four lepton types, because all observations until then could be accommodated in such a theory. In this two-generation theory, it was found that CP is automatically a symmetry for all interactions, including the weak interactions and the Higgs couplings to fermions. It is simply not possible to construct a theory with two generations of spin-1/2 particles and a single Higgs boson but no other particle types, and with the desired symmetry properties for strong, electromagnetic, and weak interactions, that does not have CP symmetry. Thus the experimental observation of CP violation proves that the two-generation Standard Model is wrong. It cannot be the complete picture of nature.

This is an interesting comment on the way physicists work. The Standard Model, in its two-generation form, was built after experiments had shown that CP was not a symmetry of nature. But no one seemed to worry about CP as they first put the ideas that together became the Standard Model. How could they ignore such a problem? The answer is that CP violation is a small effect. Physicists have learned that it is usually safe to ignore small effects and get the big picture right, and add the details that take care of the small effects at a later stage. To start with, there were too many other more critical things to check, too many predictions to test. Physicists took some time to get around to asking about CP violation. The new theory had to pass many, many tests. Only as it began to be successful did people start looking at such details.

In 1973, however, an intriguing breakthrough was made, without much fanfare, by two physicists in Japan, Makoto Kobayashi and Toshihide Maskawa. They looked at CP violation within the Standard Model. First, they proved that there can be no CP violation with only two generations. Just as in the Dirac equation, CP symmetry happened automatically in the most general version of this theory.

Then they investigated the possibility that instead of just two genera-tions of particles, there may be three. There were plausible arguments that if there are more particles in nature they should come in complete generations, namely, entire sets of particles with all properties similar to the first two generations except for their masses. But there was no argument that there should be only two, or any other number, of generations. Three was the next simplest possibility. The discussion of three generations occupies just one paragraph in Kobayashi and Maskawa's paper. But the content of that paragraph is very important. They showed that the Standard Model with three generations is not just a simple extension of the Standard Model with two generations. The two models are different in a fundamen-tal and deep way. Their symmetries are different.

Specifically, if there are three (or more) generations, then the theory does not automatically have CP symmetry, as it does with only two. (More generations give more ways to couple quarks and antiquarks to the Higgs particle, and it turns out that this is what allows some CP violation to appear.) If one wanted the model to have CP symmetry, one could impose it on the theory. But if one does not, the theory allows for CP violation in (and only in) weak interactions that are mediated by the W-plus and W-minus bosons.

At first, the work of Kobayashi and Maskawa made little impact on the physics community. It remained almost unknown to physicists in the western hemisphere. There was yet no experimental hint for the existence of a third generation. Not even all the particles of the two-generation Standard Model had been experimentally observed in 1973; most physi-cists were still skeptical about charm. The theoretical speculation that a third generation might exist did not invoke much interest; it seemed too big a step. The fact that it was needed to allow for CP violation went more or less unnoticed.

Indeed, it was not the only choice to achieve this effect. Steven Weinberg (1933– ; Nobel Prize 1979) was the physicist who first developed the Standard Model version of the theory of weak and electromagnetic inter-actions. He called it "A theory of leptons" because, in 1967, he was not quite ready to extend it to the as-yet-unconfirmed quarks. After the Standard Model theory, including the quarks, became established, he pointed out that another way to extend the two-generation theory was to add additional Higgs-type fields, and that this too could give the possibility of CP violation in the many Higgs-fermion couplings of such a theory, even with only two quark and lepton generations.

Attitudes changed rapidly with important new discoveries in 1974 and 1975, a story that was told in Chapter 10. The $J/\psi$ particle containing the charm quark was discovered, thus completing the second generation, but also the tau lepton was discovered, in data from that same experiment. Physicists soon realized that, to maintain the successful predictions of the Standard Model, the existence of the tau lepton was a strong hint that a complete third generation was waiting to be discovered. The implications of the existence of a third generation became of the highest interest. And the theoretical observation of Kobayashi and Maskawa, that CP violation is possible within the Standard Model, was perhaps the most interesting consequence of all. It began to receive the attention it deserved.

In the years since then, all the remaining particles of the third generation have been found. Furthermore, it has been shown from experiment that there are only three light neutrino types, no more. This suggests strongly that there are only three generations. So nowadays when physicists say "Standard Model" they mean the three-generation version, a theory that has been remarkably successful over thirty years of experimental probing.

We have already explained how the combination of universal weak interactions and nonuniversal Yukawa interactions translates into nonuniversal $W^{\pm}$ interactions when we examine the true vacuum state, that is, the state with nonzero Higgs field. Similarly, the combination of CP-symmetric weak interactions and CP-violating Yukawa interactions translates, in the true vacuum, into CP-violating $W^{\pm}$ interactions. The key point is that, in the three-generation theory, there are enough independent Yukawa couplings that CP symmetry is not automatic. Roughly speaking, the Higgs coupling to, say, an up quark and a top antiquark may be different from its coupling to an up antiquark and a top quark. That would introduce CP violation when we examine weak interactions of the W-boson with quarks of well-defined masses. The consequence of that is that (again, roughly speaking), for example, the coupling of the W-minus boson to a bottom quark and an up antiquark is different from the coupling of the W-plus boson to a bottom antiquark and an up quark.

The analysis of this issue is quite complicated but, when all is said and done, we find that, in the three-generation Standard Model, the Higgs

Yukawa couplings introduce not only the quark masses but also one (and only one) CP-breaking parameter into the theory. There is a rather technical counting of parameters that tells us there is no such parameter for two generations and only one with three generations. This single CP-breaking source appears as a complex phase in some of the quark weak couplings; naturally it is called the Kobayashi–Maskawa phase.

When we say that a model has a number of parameters, we mean that the energy function that includes all terms allowed by the symmetries has this number of coefficients. The model and its symmetries only tell us what are the various combinations of fields that can appear, they do not tell us the numerical values of the various coefficients. One needs experimental information, that is, a measurement of a rate that depends on a particular term, in order to determine the numerical value of the corresponding coefficient. Thus, if a model has a certain number of parameters, one needs, in principle, to perform precisely this number of (appropriately chosen) experiments in order to know the values of all parameters. Then one can build on this knowledge and predict the results of all further measurements. We have already said that, for example, the strong interactions depend on a single parameter. That means that, in principle, we need to measure the rate of a *single* strong interaction process. That would allow us to determine the value of the strong interaction coupling and predict the rates of all other strong interaction processes.

The fact that the Standard Model has a single CP-violating parameter is then very significant. It means that we need to measure the rate of a single process that breaks CP. That would determine the size of this CP-breaking parameter, the so-called Kobayashi–Maskawa phase, and allow us to predict the size of all other CP violations. If we measure a second effect that is different between particles and their antiparticles, we can compare that to our prediction. If the prediction were right, we would become more confident that the Standard Model way of breaking CP is indeed the one realized in nature. If our prediction failed, the Standard Model would be proven wrong. Thus, as concerns CP violation, a second (and third and fourth . . . ) measurement will constitute a *test*.

We have already mentioned the first observation of CP being violated in 1964, by Cronin, Fitch, and their collaborators. With the three-generation Standard Model at hand, physicists could use the result of this experiment to fix the value of the single CP-breaking parameter of this theory, that is, the Kobayashi–Maskawa phase. But to test it with a second precise measurement, they had to wait for more than thirty years. We will tell this story later. But now, let us ask how all the things that we have learned about the Standard Model theory and, in particular, the fact that it breaks CP, affect the history of matter and antimatter in the Universe.

# 15

# *IT STILL DOESN'T WORK!*

## Running the Clock Forward: The Standard Model

Armed with modern particle physics or, more concretely, with the Standard Model, to enrich our understanding, how does the history of matter and antimatter look now? In our previous run through the history of the Universe, we suggested a possibility known as baryogenesis. Does the story work if we use the Standard Model to set the physics parameters?

Baryogenesis starts from equal amounts of matter and antimatter. The imbalance between them develops as a result of CP-violating interactions. The small difference in the physics of matter and that of antimatter determines how much matter would be there in the present Universe, where all antimatter has disappeared. But does it work? Do we get *the right* number density of baryons, compared to that of photons? In a sense this number constitutes a measurement from the Universe of the strength of CP-violating process—does it match the CP violation we now know is there in the Standard Model? Can we explain the observed baryon asymmetry of the Universe with the Kobayashi–Maskawa mechanism?

First, we must ask whether all three of Sakharov's conditions are met by the Standard Model. Most important for our purposes, the model has CP-violating weak interactions, with only a single source of CP-symmetry breaking, the Kobayashi–Maskawa phase.

The model also has baryon-number-violating interactions. This fact was first realized by Gerardus 't Hooft (1946– ; Nobel prize 1999).

Moreover, the rate of processes that change baryon number depends on the temperature in a way that, when we run the clock forward, fits nicely with what is needed for baryogenesis. At early times of the Universe, prior to about $10^{-11}$ seconds, these interactions were active, that is, they had a comparable rate to other processes. At this time a phase transition occurs, the value of the Higgs field in the Universe goes from being on average zero, to being everywhere constant but nonzero. So before this transition the particles have zero mass, but after it they get masses from their interactions with the constant Higgs field. In particular, the W- and Z-bosons that are the mediators of the weak interactions get very large masses, of the order of 100 times the proton mass. These masses suppress the baryon-number-changing processes of the Standard Model almost completely. Consequently, any baryon number that exists after this transition is completed will persist essentially unchanged until today.

A reminder about our use of the term "baryon number" in this story may be helpful—what we are really counting here is quarks minus antiquarks. The events that we are describing took place in the very early Universe. When the Universe was younger than about a millionth of a second, quarks and antiquarks were effectively free particles, not formed into definite hadron combinations. The soup of quarks and antiquarks filling the Universe at this time was so dense that the typical distance between neighboring particles was small compared to the typical size of an isolated hadron. Only after the Universe had expanded and cooled till the temperature dropped below about $10^{13}$ kelvin were quarks confined into separated baryons (protons and neutrons) and antibaryons (antiprotons and antibaryons) and mesons. However when counting particles minus antiparticles physicists use the term baryon number even in the earlier stage when there are really no distinguishable baryons. When the Universe is young enough that (anti)quarks are free, the baryon number is simply one-third of the quark number, the number of quarks minus that of antiquarks.

The next of Sakharov's conditions, lack of thermal equilibrium, is also met in this picture. The transition from zero Higgs field throughout space to a space filled with a fixed, constant, different-from-zero Higgs field is itself a possible source of the period when thermal equilibrium does not apply. So let us form a scenario for what happens in this transition. The picture is similar to the phase transition that occurs when a superheated liquid becomes a gas. In the liquid some region becomes a bubble of gas. Because the conditions are such that the gas phase is actually the physically favored phase, any big enough bubble once formed begins to grow, until eventually all bubbles meet and the entire volume is gas rather than liquid. In the Universe the bubble is a region in which the Higgs field is nonzero, while the outside region has zero Higgs field value.

The skin of the bubble is a region where the Higgs field value changes rapidly over a short distance. The analogy is only a crude one; one major difference is that, in an expanding Universe, the space between the bubbles also grows, so fast that bubbles never meet. So, in this picture, the entire currently observable Universe is the inside of a single bubble that formed about $10^{-11}$ seconds after the Big Bang.

Now what happens as the transition occurs? Imagine that a bubble forms and begins to grow. All our attention has to be on what happens at the surface of the bubble, which sweeps through space at great speed. Imagine this surface has a definite thickness, the region over which the value of the Higgs field is changing from the zero value outside the bubble to the large value it has inside the bubble (and everywhere in the Universe today). The particles and antiparticles interact with this bubble wall. Some of them are swept through it as it passes and end up inside the bubble. But others bounce off and end up still outside it.

Now suppose that the CP-symmetry breaking is such that the probability for a quark to get through the wall is a little greater than that for an antiquark. Then inside the bubble one would have an excess of quarks over antiquarks. Furthermore, because the particles are massive inside the bubble, the processes which can change baryon number are much suppressed there, and so any excess persists. However, outside the bubble baryon-number-changing processes are still active, so any excess of anti-quarks that begins to build up there is removed by these processes, which continue to drive the system there toward thermal equilibrium, and thus maintain equality between quarks and antiquarks in this region.

Thus all three of Sakharov's conditions can be met in the Standard Model. In particular, the CP violation in the Standard Model does indeed give differential transmission through such a bubble wall. All that remains to be done is to calculate the numerical consequences of this effect. The calculation is complicated and we will not describe its details here. Moreover, the precise result depends on many details, such as the form of the bubble surface. That, and indeed the very formation of such a bubble in the first place, in turn depend on quantities we do not yet know well, such as the mass of the Higgs boson. But one result is clear: any baryon asymmetry that develops inside the bubble with any choice of Standard Model parameters is *much* too small. The Standard Model could lead to the right qualitative picture: no antimatter and little matter. But the quantitative picture is disastrous.

There are two reasons that the Standard Model fails to explain the baryon asymmetry of the Universe with this calculation. The first problem is that, as best we can tell, the parameters of the bubble wall are not right. As the experimental lower limit on the value of the Higgs mass goes up, the picture gets worse and worse. The story we have just described corresponds to a sudden phase transition, with a well-defined bubble surface. The current values of the possible Higgs parameters in the Standard Model (with only a single Higgs field) do not give such a sharp transition. So the simplest Standard Model just does not give the picture we described above. We can however patch that up, at the cost of adding something, such as more Higgs-type particles, to the theory. Then we can fit all the current data and still have a clean phase transition with a well-defined bubble wall.

But even if we fix this point, and there is such a phase transition that is as sudden as we would like, the Standard Model cannot give enough matter. The problem has to do with the amount of CP violation. The number of baryons we would calculate in the present Universe is smaller by at least sixteen orders of magnitude than the actual observed number. Now we said our calculation had many uncertainties, but it is not that uncertain. No amount of stretching the details can contrive to make this number big enough. If the Standard Model were the whole story, we could not have clusters of galaxies, and not even our own Milky Way in the Universe. The number of protons and neutrons could at best fit the number contained in a single cluster of stars.

The Kobayashi–Maskawa mechanism for CP violation depends on a number of details of the Standard Model. If the $W$ coupling to any of the nine quark–antiquark pairs were zero, there would be no CP violation. If any two down-type quark masses were equal, there would be no CP violation. If any two of the up-type quark masses were equal, there would be no CP violation. While none of these conditions are met, none of them are very far from being true. Thus, it turns out that each of these conditions takes a toll on the amount of baryon asymmetry generated at the bubble wall. The mathematical details are such that the ratio between the baryon number and the number of photons cannot be large enough, independent of the details of the phase transition. It is just impossible for the Kobayashi–Maskawa mechanism to produce baryon asymmetry as large as observed.

So the mystery of the missing antimatter is back. Only now it is a different puzzle. Now we understand how a baryon asymmetry can be produced in principle by a consistent history, involving baryon number violation, departure from thermal equilibrium, and CP violation. But the current theory of particle physics, that is, the Standard Model, fails to give the required result. We can even identify the ingredients of the theory that failed: the phase transition is not right, and CP violation is much too small. Both failures are dramatic. We conclude that the Standard Model, as we have described it so far, is incomplete. There must be new sources of CP violation, different from and, in particular, much less suppressed than the Kobayashi–Maskawa mechanism. The Higgs particle needs some help; perhaps more than one Higgs-type particle is needed, to get the right kind of phase transition and bubble wall.

## Now What?

Baryogenesis tests the Standard Model picture of CP violation. The Standard Model fails this test. Therefore, baryogenesis tells us that the Standard Model is not the full story. Now we must look again at the clues to find a new answer to our mystery.

Our detective work would be easiest if indeed baryogenesis happened at a time close to $10^{-11}$ seconds—the time of the electroweak phase transition. Perhaps nature used a bubble wall similar to the one that we described in our failed attempt at Standard Model baryogenesis, but has new sources of CP violation to do the job. There are many ways to extend the Standard Model to get the right phase transition properties and a bigger baryon asymmetry. If that is the right answer, we may be able to directly observe these new interactions that violate CP in our laboratories. Some of the relevant experiments are operating as this book is being written; we will get to them soon.

If the baryon asymmetry was produced at much earlier times (and, equivalently, much higher temperatures) then it will be much harder to prove (or to disprove) any theory of how it occurred. The relevant processes then must involve the exchange of extremely heavy particles. Our accelerators may never have enough energy to produce such particles. There will be no direct evidence for their existence. We can, however, hope to achieve some circumstantial evidence, and to do so we need a full theory for the processes that drive the asymmetry. Such theories certainly involve physics beyond the Standard Model.

So whether the answer is in modification of the CP violation involving quarks, or in some other realm, we now know that it involves physics that is beyond the Standard Model. How do we look for physics beyond the Standard Model? To understand some aspects of such a scientific adventure, we go back to the fifteenth century.

Imagine a wrong model of Earth, a possible picture as far as most Europeans knew, before the journey of Columbus. It is a ball with a surface that has only one big body of land, that is, Europe-Asia-Africa, surrounded by water. Compared to the then-standard model of a flat earth, the model of a sphere with one continent was a progressive idea in the fifteenth century. What happens if you take this idea seriously? It allows you to make predictions. If it were correct, then you would be able to travel from Spain to India in three very different ways.

Two of these ways were known and difficult—the land route to the east, and the sea route to the east, which required sailing around Cape Horn. But the model suggests that a third route might perhaps be easier—a sea route to the west. Columbus was the first to try this third way. His journey, based on the wrong but progressive model, led to a fantastic and, to him, quite unexpected discovery: a new continent. It showed that the model of one big body of land is wrong. It was a breakthrough in European knowledge of Earth, and this knowledge changed history. (Probably some Vikings knew it much earlier, but their information apparently had not made it to Spain or Italy. The fact that Eratosthenes had made a quite accurate measurement of the radius of the earth in the third century B.C., and his results, were clearly not well known in Europe in the fifteenth century.)

Like Columbus's great discovery, any discovery of an inconsistency with the prevailing theory, any observation that does not fit the current Standard Model, would be a major achievement. Such a discovery always opens up new territories to explore. Often the search, like Columbus's voyage of discovery, is guided by a new theory or model for the untested regime, ideas which, like the model Columbus used, may turn out to be partially right but also wrong in some important aspects.

Even when a theory is well established in some regime of conditions, physicists hope to find such a conflict every time that they suggest or perform an experiment that explores a new regime. What is new could be in the level of precision, or in the fact that the experiment searches previously uncharted ranges of collision energy or types of colliding objects. Why do we hope to find conflicts with our predictions? One usually learns more when the prevailing theory is found to fail than when it is found to encompass the new regime in addition to those where it has previously been tested.

This makes physicists contrarians: there is nothing that makes the physics community more excited than the failure of previously successful theories that it created. Such a failure tells us not that the theory is entirely wrong, but that we have found a place where it is incomplete. We know we have missed some important detail. If we find inconsistency with the Standard Model, it will to lead us eventually to better and deeper understanding of nature. Experimenters searching for CP violation hope that their results will eventually be inconsistent with the Standard Model predictions and open a window to new physics.

Now, imagine that America was not there. Columbus might then have found his way to India, as he hoped to do. Such a journey would still have improved the European understanding of Earth. Then his new theory would have been validated and, moreover, he would have learned the values of some of the important parameters in that theory, such as the circumference of the Earth, which were previously unknown to him.

Likewise, even if our experiments do not exclude the Standard Model picture of CP violation, each new result can have great value because it allows us to refine our theory, defining it better. If the violation of

CP measured in new processes is consistent with the Standard Model predictions, its measurement will improve our understanding of this model and refine the determination of its parameters. It will also suggest further, more precise tests via the measurements of yet other processes.

Columbus sailed on the best ships that he could get, loaded them with as many supplies as possible, and went west. He did not have absolute confidence that his ships and supplies could get him where he wanted. But he still left his home harbor.

Similarly, scientists seek the best way to make new discoveries. They would like to search for large CP-violating effects that, once measured, can be interpreted in a clear theoretical way. They would like their experiment to collect as many data as possible. But in the end, they do what they can do. If we can only measure processes where the CP-violating effects are not large and where the theoretical predictions are not so very clean, we still want to measure these processes. Whatever the result is, with combined experimental and theoretical effort, we can always learn something, sometimes a small step, sometimes a larger one, to bring us a little closer to the understanding that we seek.

Of course, the ship that allowed Columbus to explore was critical to his discoveries. So with science, the questions we can explore with experiment depend on the equipment we can use to do the experiment. The advancement of basic physics knowledge goes hand in hand with advancements in the technology of accelerators, of detectors, and of data-handling capability. New developments in each of these areas are essential for progress. These developments are sometimes driven by the physicists' desire to make certain experiments. Sometimes they come from developments in other areas, such as new possibilities in electronics, which find application in high-energy physics experiments.

Developments of the physics world often find application elsewhere as well. All medical X-ray machines are powered by small electron accelerators based on the design for an electron accelerator developed at Stanford in the early 1950s, much the same basic design as that of the two-mile SLAC accelerator that has already played a role in our story several times, and will again. More recently, particle physicists at CERN, the European particle physics laboratory, developed a way to share data and information over the Internet, among widely dispersed members of their experimental collaborations. They named their invention the World Wide Web, and published their protocols for all to use. This invention, soon improved by the development of user-friendly browsers, has changed the way the world shares information. These are perhaps two of the largest impacts, but there have been many such spin-off effects from high-energy physics research. When scientists push the limits of technology to undertake a challenging task they may create new options for less esoteric uses as well.

# 16 TOOLS OF THE TRADE

## Accelerators

The history of particle physics we have described throughout this book is based, to a large extent, on the invention of accelerators as tools to probe matter structure at subnuclear scales. In this chapter we describe what these machines are like, particularly the modern versions that are now testing the Standard Model picture of CP violation in great detail.

You can think of an accelerator as a gigantic microscope, allowing physicists to view phenomena at ever smaller scales as the energy of the accelerated particles is increased. To understand this point you need to know two facts, namely, that particles behave like waves and that the wavelength of the associated wave gets shorter as the particle's momentum (and thus also its energy) gets larger.

Now think about sitting on a pier and watching the waves pass under it (figure 16.1). The pilings of the pier are small compared to the wavelength of a wave, and they scarcely disturb it at all. But when the waves crash into a seawall the wave is scattered to pieces (figure 16.2). The length of the seawall is large compared to the wavelength. Any wave can only be disturbed by objects that are large compared to its wavelength. We "see" things by the disturbance they cause in a wave, by the way the wave, be it light or a particle beam, is scattered off them. We then use what we know about the behavior of the probing wave to determine the shape and structure of the thing that caused the disturbance.

Fig. 16.1 Sitting on a pier and watching the waves pass under it.

We do not see atomic-scale structure with visible light, but we can "see" it when we use higher-energy photons, in the ultraviolet to X-ray regime. Likewise we needed high-energy electrons to "see" the quark structure inside protons and neutrons. Accelerators with even higher energy, such as that now under construction at CERN, can be expected to yield new discoveries because they are able to see things at scales that have never been probed before.

A high-energy accelerator also gives us access to a second new regime, not just smaller distance scales, but also higher mass scales. We cannot produce something from nothing; the mass of a produced particle must come from conversion of energy from some other form to this one. To make a massive particle such as a Z-boson we must cause a particle collision with high enough available energy. Often we can infer the existence of massive particles by their role in the theory and by their indirect effects (particles present in unobserved intermediate stages of a quantum process). Eventually we can be sure the theory is correct when we actually produce the predicted particles.

Fig. 16.2 The waves crash into a seawall.

The basic principles of accelerator physics are quite simple. One accelerates charged particles by using the fact that electric charges feel a force due to electric fields. In truth, it would be better to say that one energizes the particles in this way. For example, after only a few feet along the two-mile SLAC accelerator the particles are traveling at $0.99c$ ($c$ is the speed of light). By the end of the two miles they have sped up by only about 1% more, to $0.999999999999c$. But their energy has increased by a factor of almost a million. For particles traveling close to the speed of light, as Einstein's special theory of relativity tells us, a small change in speed corresponds to a very large change in energy.

The second basic idea is that one can steer and focus a beam of energetic charged particles using magnetic fields, due to the fact that a charged particle moving in a magnetic field feels a force that is perpendicular to its direction of motion at that instant. Such a force does not change the energy of a particle in the beam at all, but it changes its direction of travel.

Of course, this simple description ignores all the complexity needed to achieve these desired effects. The engineering and technology demands of a modern accelerator facility are far from simple. Inside the accelerator one must maintain a pressure less than a millionth of a millionth of atmospheric pressure. Intense electromagnetic fields must be produced, varied in time and sculpted in space very precisely to give exactly the desired acceleration and steering. Coordination between one part of the facility and another requires high-precision monitoring and feedback control systems.

The design, building, and operation of such machines constitute a research area in itself, with its own set of theorists and of practitioners. Even after new ideas are developed and tested in small-scale experiments, each new machine requires a multiyear project of research and development before a final design is achieved. Typically several more years are required to build the facility, and so of course one wants to design it so that it can be expected to have many years of useful and reliable operation.

Just a few details will show how tricky a task the engineers must perform. A modern storage ring is hundreds of meters in circumference, and its inner region must be evacuated to a pressure of about one in a million millions of atmospheric pressure, or else collisions of the beam particles with the gas in the pipe would totally destroy the capability to circulate closely packed bunches of electrons or positrons. Every joint or flange in the structure must provide this level of seal, and the components must be made with clean-room techniques to ensure that no surface has so much as a fingerprint to be a source of outgassing of atoms into the vacuum region.

The beam pipe is a metal structure so every time a bunch of electrons pass through a section of it they cause currents to flow in the walls of the pipe, which in turn cause electromagnetic field effects inside the pipe. These fields can disturb the next bunch of particles to come by. Only with very careful monitoring, and controls with very fast feedback capability, can one achieve a stable situation with dozens of bunches circulating in the ring—and one needs as many as possible to get the rate of collision events high.

Then one must design the beam steering, to get the electron bunches from one ring to meet with the positron bunches from the other ring, and to get the bunches back in their respective rings after they pass through one another (only a small fraction of the electrons and positrons in each bunch actually interact and disappear in one beam crossing). The beams are steered with magnetic fields, but when their path is bent they radiate. The radiation must be absorbed and not enter the detector or else it could destroy the detector components and swamp the interesting information with spurious signals. It is intense enough to melt components unless they are properly shielded and cooled. Fortunately this intense radiation is short-lived and disappears shortly after the machine is turned off, but clearly, with such high voltages and high-radiation areas, careful attention to safety controls is an important part of accelerator operation.

Once such an accelerator is built, even one designed specifically for a particular set of experiments, the processes of interest to physicists today occur only rarely. Perhaps one in a million collision events produces the interesting process, even when the accelerator has been designed to give the best possible conditions for it. We generally need to collect hundreds of events of a given type in order to have a precision measurement, to know whether we are seeing them at the rate expected by our theory. So once an accelerator is built it must operate reliably twenty-four hours a day for months at a stretch, to produce the millions of collisions needed for any single study. That too places severe demands on its design, as every component must be designed for high reliability. The larger the facility, the more stringent this requirement becomes.

## Detectors

An accelerator produces particle beams and sets up the collisions between them, colliding particles in the beam either with a target such as a tank of gas, or with particles in another beam. Physicists call the former *fixed target* experiments and the latter colliding beam or *collider* experiments. But the collisions are not the end of the story. Specialized detectors must be built to be able to decode what happened in the collision. This brings a whole new set of specialists into the game, the experimental physicists who design, build, and operate the complex set of instruments, collectively called a detector, that serve this purpose. (We recognize that the term experimental physicist is poorly constructed; it is the physics that is experimental, not the people. We ourselves are called theoretical physicists, which is even worse. We certainly hope the word theoretical modifies the physics that we do and not our status as physicists. But these are the short-hand terms used within the field.)

In modern particle physics jargon, a detector is actually a multicomponent layered instrument, with each layer designed for a specific task. In a collider experiment the detector facility is a cylinder surrounding the collision region. The layers are typically designed as follows: Innermost is a precision tracking device, next a larger-volume tracking chamber, and then a layer that measures charged-particle velocities. Outside that one has a *calorimeter* (which is what physicists call an energy-measuring system since a calorie is a unit of energy). All of that is typically inside the coil of a large cylindrical electromagnet, which induces a strong magnetic field parallel to the axis of the cylinder throughout this core region. Outside the magnet coil comes a layer that completes the magnet and the detector system by providing a large region of iron. This keeps the magnetic field outside the coil within a restricted region. This iron is interleaved with detector layers. These coarse-grained tracking detectors record the passage of any charged particle that manages to penetrate through all the inner layers to reach this region.

The innermost precision tracking system is called the *vertex detector*. Its principal purpose is to reconstruct the tracks of charged particles accurately enough to determine whether they all emerge from the beam-beam collision region, and to find those sets of tracks that do not, but instead emerge from a point, or vertex, some hundred or so microns away from that region. These vertices are the key signature that a B-meson has been produced. B-mesons and the baryons that contain b-quarks have a long enough half-life that, when produced moving at close to the speed of light, they travel a few hundred microns, and so one can actually image their decay in this way. Thus the vertex detector is a key component if you want to study these particles. Without it we could not readily tell that a B-meson had been produced; the vertex detector allows us to pick this particular needle out of the haystack of hadronic events.

The vertex detector is a multiple-layered set of silicon chips, each layer divided into many pixels. Each pixel registers when a charged particle passes through it. The track is reconstructed by connecting the dots from one layer to the next. This task can be achieved very rapidly, so the detector system can be programmed to record events where a B-meson-like vertex is seen, and not to keep a record of others that do not fit this criterion.

The second tracking layer is less precise but has larger volume. This layer is designed to be able to measure the curvature of every charged-particle track in the magnetic field, and thereby to determine the momentum of the particle that produced that track.

The tracks of the highest-energy particles bend very slightly, so a relatively large region is needed over which to detect and measure this curvature. It is simply too expensive (as yet) to instrument a large enough region for this measurement with precision silicon tracking. Instead the device most commonly used is one called a *drift chamber*.

A drift chamber has hundreds of very fine wires stretched along it in a complex pattern. The wires form a grid that effectively divides the chamber into Geiger-counter-like regions. The chamber is filled with a readily ionized gas and certain of the wires, known as the sense wires, are held at a high voltage compared to those around them. As a charged particle moves through the gas it ionizes gas atoms along its path. The electrons that are liberated in this process drift toward the sense wires. Instrumentation on the end of the wires registers which wires the electrons strike. The pattern of wires is such that this allows a three-dimensional picture of the charged-particle tracks to be reconstructed.

The next layer is a velocity-measuring layer, also called the *particle identification* system. Once we know both the momentum and the velocity of a given particle we can determine its mass. If that is done accurately enough we then know what type of particle it is. The trick to measuring velocity is an effect called Čerenkov radiation, named for the physicist and detector developer Pavel Čerenkov (1904–1990; Nobel Prize 1958) who first recognized the power of this effect. The effect is an interesting one; it is a shock wave of light that occurs when a particle travels through a medium faster than the speed of light in that medium.

All particles travel slower than *c*, the speed of light in a vacuum. But in a medium the effective speed of light is slowed by the scattering of the light on the charged particles in matter. So a particle moving at close to *c* can travel through a medium faster than a light wave does. Then, just as a boat in water makes a bow wave with an angle that depends on its speed, or a plane traveling above the speed of sound makes a sound shock wave, so the particle makes a light shock wave in the medium. The opening angle of the cone of light that surrounds the particle's trajectory can be measured and used to determine the particle's speed.

The next layer is the calorimeter or energy-measuring device. The technology chosen for this layer varies from one experiment to the next. Later we will describe experiments at facilities called B factories. In these experiments the calorimeter is made from large crystals of cesium iodide. These crystals absorb energy from almost all particles, charged or neutral, that enter them and reradiate it as a light pulse that can be detected and converted to an electronic pulse by a photomultiplier device placed at the base of the crystal (the base being at the outside of the cylinder of crystals). The crystals are long and narrow, long enough that most of the particles are stopped in them, and narrow enough that they provide good definition of where the particle arrived that deposited energy in them. A single particle illuminates one or a few neighboring crystals depending on its direction of travel. The crystals all point toward the beam-beam interaction point.

Next comes the magnet coil that provides the magnetic field threading all the inner layers of the detector. Finally the outer layer is outside the coil. This layer contains another set of particle-tracking devices; typically these are interleaved in the iron of the magnet flux return, a large amount of iron needed to control and contain the magnetic fields outside the magnet coil. The outer tracking devices must cover a large area but do not need high precision. There are many different technologies possible here. The task is again to recognize and locate the passage of charged particles. Most particles are stopped in the calorimeter layer. Only one type of particle, namely, muons, almost always reaches the outer tracking layers, so this set of detectors is often referred to as a *muon identification* system. Roughly speaking, any track that goes through this outer layer (and deposits very little energy in the calorimeter) can be assumed to be that of a muon. This is an important extra piece of information, as the pion and the muon happen to have very similar masses, so one cannot distinguish them by knowing both momentum and velocity; but pions will be stopped by the calorimeter layer. Muons will pass right through it, and through the magnet coil, and on out through the outermost layers.

Early particle detectors, such as cloud chambers, emulsions, and bubble chambers, were particle-tracking devices where the tracks were recorded photographically. Then the pictures had to be visually scanned to find the patterns of interest. High-energy physicists in those times were mostly men, but they employed large numbers of women as scanners going through the record with care to find track patterns that signified the type of events that the physicists were looking for. The transition from these types of tracking devices to those described above occurred with the onset of large-scale computation capability. The signal processing that turns the individual pixel counts and pulses on wires in the drift chamber, signals of energy deposited in the calorimeter, and so on, into pictures of the event is part of the software that each experiment must develop.

Pictures can indeed be made by the computers, but only because these provide physicists with a valuable tool to check that everything is functioning as it should. No one looks at pictures of most of the events. They are all processed and interpreted by the computers. The transition from visual to computer particle tracking was a key to the ability to process the huge amount of data needed in modern experiments, such as the B-factory experiments that probe CP violation.

The general elements described above are common to all collider detectors, but the details of each one are different. Designing such a detector involves a series of decisions, each a compromise to achieve the most capability within the cost and space constraints. Clearly the detector components must be built into some supporting structure, and the utilities to power them, keep them cool, and carry their signals out to a recording system of computers must be designed into the structure as well. The design and construction of such a detector must proceed alongside and be integrated with that of the accelerator that will provide the collisions that the detector will view and record.

Typically the team of physicists and engineers working on the detector are a different group from those working on the machine, but strong liaison between the two teams is essential. This dual-team approach continues once the experiment begins to operate, with one team concerned with the operation of the accelerator while the second team operates the detector and oversees the data collection as well as the data analysis projects that sort through those data to extract particular physics results. This second team is much larger. At each high-energy experiment it is an international organization which involves hundreds of physicists.

## Data Handling and Analysis

The best detector in the world is useless if we cannot record and analyze the data it produces. In modern high-energy physics experiments, that is a significant challenge, as the amount of data that must be collected to make the measurements of interest to us is enormous. For example, at the B factories, we want to study particular B-meson decays that occur at a probability as low as one in a million, and we need hundreds of events of a particular type to have even 10% accuracy in measuring this probability. That means we need to produce many millions of pairs of B-mesons and anti-B-mesons.

This is the reason for the term B factory—we really need a reliable production line to achieve that feat, and then also the patience (and funding) to continue to operate it for many years. Then, as it operates, we must collect, refine, and store the data. Many different studies based on these data must be performed, so both immediate analysis and subsequent reanalysis must be possible.

Alongside that, we need large-scale simulations of the experiment which model what should happen in it according to the theoretical predictions. The comparison of the real data with the simulated data is the only way we can tell whether our theory is (or is not) correctly describing what occurs. We need to generate as much or more simulated data as real data so that statistically meaningful comparisons of theory and experiment are possible. We have to simulate all the processes that could happen, not just the ones we are interested in studying. It takes a major effort to eliminate the possibility that a small fraction of the much more frequent uninteresting processes could look enough like the rare process that we are searching for to fool us. We must show by simulations that we have a method of analysis that can discriminate well enough to ensure that all such fake or *background* events contribute at most a small correction or uncertainty to our measurement.

All this requires not just the computer facilities but also specialized software, which must be developed in conjunction with the detector design. Each element of the detector produces an electronic record of each collision event that occurs. It takes some time to read out all this information from the detector into some data storage system. While the readout is occurring, the detector is effectively turned off, so subsequent collisions that happen during this process are not recorded. Hence one cannot write out the information for every single collision event that occurs. The ones that are interesting are quite rare but the job of picking these needles out of the haystack of less interesting events (that is, of already well-understood processes) must begin at the first stage of data recording. One must develop criteria that trigger the decision to record an event, criteria that can be checked very rapidly and that are designed to separate the needles from the straw with high efficiency.

Even so the data stream from such an experiment is enormous; for example, in each B-factory experiment over a terabyte of data is collected in a month of running. These need to be recorded both in raw form and in a form where the software has converted the set of electronic hits into a map of tracks and the corresponding particle identifications, and to be made accessible for searches by physicists who wish to analyze these data, all over the world. This part of the design of the project is as critical as the design of the detector itself, and the two must develop hand in hand.

## How Projects Develop

It takes a team of hundreds of physicists to design, build, and maintain such an instrument. The process begins with a small group with the idea that a particular type of facility will provide interesting results. They produce an initial idealized design, idealized in the sense that every component is assumed to work perfectly and no engineering issues are considered, except for the obvious fact that you cannot put two different detector elements in the same place. Then some crude simulations are developed for experiments with such a detector.

This stage is critical, to study whether it is possible to build a detector capable of extracting the physics results of interest. One can use this model to investigate what choices can be made to improve one or another feature, at the price of perhaps degrading something else. These trade-offs, as well as cost considerations, which force other trade-offs, lead eventually to a basic optimized design.

On the basis of this preliminary work, a project may receive some funding for a design development stage. Test versions of critical parts of the detector, particularly any novel element of it, need to be built to ensure that the parts work as designed. Engineering studies need to be made to ensure that this facility can actually stand up, or be opened up for servicing, or run without overheating, and so on. The project begins to have a certain momentum and the collaboration begins to grow to do all this work, but the facility is still tentative. The project may already have been chugging along for a few years at this stage, and if everything looks good it is time to move to the next stage.

The critical step for a project is to obtain funding and approval to proceed to building the experiment. A B factory, for example, is a multimillion dollar project, and its construction takes a few more years. To obtain approval, a detailed proposal must be written, with a full technical description of the design and its physics capabilities. If the project is an international one, as most are today, all the relevant countries must make decisions to support it, and agreements specifying who will do what, and how it will be paid for, must be spelled out. You can imagine the elation in a collaboration when approval is received, and the intense frustration of those whose proposal for a project is not accepted.

Once a project is approved it will take a few more years to build it. Then it must be operated reliably for several more years to obtain all the results it can produce. Perhaps the first new results will appear much sooner than that, but a project of this scale is not undertaken unless it has many possible measurements to make. All of this may give you some idea why modern particle physics experiments involve collaborations of hundreds of physicists and take place over many years.

# 17
## SEARCHING FOR CLUES

## Where Are We Now?

The observed baryon asymmetry of the Universe suggests that the Standard Model does not provide the full picture of CP violation. So we look for new interactions that violate CP. To make real progress, we need experiments that can discover these new sources of CP violation. How can we attempt to do that? There are two different approaches to try to find new sources of CP violation, effects not included in the current Standard Model. First, we can measure processes where the Standard Model predicts that there are no CP violating effects or that these effects are very, very small. Then, if significant CP violation is observed, it clearly signals new sources of CP violation. The second way is to look for processes where the Standard Model gives clean predictions for large CP-violating effects. Then, if the measured CP violation is either larger or smaller than these predictions, we must be detecting new sources of CP violation, since new effects can either add to or cancel Standard Model effects. And, if the measurement falls within the predicted range, we can then refine the Standard Model through a better determination of its parameters, including the one CP-violating parameter, the Kobayashi–Maskawa phase.

An example of the first approach is the search for a particular type of photon emission from a neutron. You may think that, since a neutron has no electric charge, it cannot emit a photon. But in fact, since it is composed of charged quarks, there are small effects which depend on the distribution of these quarks within the neutron. The effect we are looking for, a particular pattern of photon emission relative to the direction of the neutron's spin, would be forbidden if CP were an exact symmetry. In physicists' language, we are searching for the electric dipole moment of the neutron.

The Standard Model predicts a tiny electric dipole moment of the neutron, about a million times smaller than our best current experiments can measure. But many extensions of the Standard Model predict that there is a much larger effect, within the reach of the next generation of experiments. Already the currently measured upper limit on the size of this effect has eliminated many ideas for extending the Standard Model—they predicted too large an effect.

However, there are still ideas for models, extensions of the Standard Model, in which new effects give enough CP violation to explain the baryon asymmetry of the Universe, but give a neutron dipole moment of the neutron that is smaller than the current limits—but not by much. So far, however, the very beautiful experiments that searched for the electric dipole moment of the neutron did not find any signal. The search goes on. If one day a signal is found, it may mean that the two ends, the particle physicists' quest for understanding CP violation and the cosmologists' quest to understand baryogenesis, have met. We will be more than happy to rewrite our story!

To exploit the second approach we need experiments at accelerators. Within the Standard Model, large CP-violating effects can appear only in processes that are dominated by weak interactions. If we further focus our attention on observables that are cleanly predicted, we are led to consider decays of neutral mesons. More specifically, only certain very rare decays of neutral K-mesons and some rare (but not quite so rare) decays of neutral B-mesons satisfy the two conditions of large CP-violating effect and clean prediction. Of these, the first to be measured are the B-meson decays. But other experiments may pursue the rare K-meson decays, because the relationship between the two sets of results will also be useful. If we find that the different approaches cannot all be fitted by a single choice of Standard Model parameters, this will be a sign of some new physics effects. It will take more than one result to pin down just what such effects might be.

## Testing the Standard Model in B-Meson Decays

In the early 1980s, Ichiro (Tony) Sanda suggested a new way to study CP violation, beyond the known effects in K-meson decays. Sanda was then a young researcher at the Rockefeller University (now he is a professor at Nagoya University). He pointed out that, like the two neutral K-mesons, K-zero ($K^0$) and anti-K-zero ($\overline{K}^0$), there are pairs of neutral B-mesons, B-zero ($B^0$) and anti-B-zero ($\overline{B}^0$), that are particle-antiparticle mirrors of each other and that these systems should provide an excellent laboratory for studying particle–antiparticle asymmetries, that is, CP violation. Sanda and his collaborators suggested that a measurement of a CP asymmetry in the decays of these neutral B-mesons might provide the long-sought-for test of the Kobayashi–Maskawa mechanism of violating CP symmetry.

Sanda expanded on this idea with a number of different collaborators, including Ikaros Bigi (now a professor at Notre Dame University). Bigi and Sanda reviewed possible decay of neutral B-mesons that would be interesting to study, such as decay to the two-particle state comprised of a J/psi meson ($J/\psi$) and a K-short meson ($K_S$).

(Remember, the J/psi is a single particle, made of a charm quark and matching antiquark—its double name comes from the fact that two separate groups found it.) This decay possibility is one of a long list Bigi and Sanda suggested for study in their 1981 paper. Today this particular decay mode is often called the "golden mode" by physicists working on B-meson decay experiments. It looks good both to experimenters and to theorists. It is readily identified experimentally (though very rare) and it has a very clear prediction for what should be observed if the Standard Model is correct.

Helen says: "I have now spent many years of my life working on the physics of B-meson decays, but I remember my negative reaction when I heard an early seminar on the subject, probably around 1982. Ikaros Bigi had been a postdoctoral researcher at SLAC and then moved back to Munich (where he had studied for his Ph.D.). He came to visit SLAC and talked enthusiastically on his ideas about B-meson decays. I was less enthused; I thought to myself that this pretty idea could not possibly be tested. How wrong I was! What neither I nor anyone else knew at that time was that particles containing bottom quarks (such as the B-mesons) have even more suppressed weak decays than those containing strange quarks. This gives B-mesons a much longer half-life than anyone at that time expected. The long lifetime makes the current experiments possible, because it means that B-mesons can travel far enough for their decays to be detected with new precision tracking detectors. The invention of charge-coupled device (CCD) chips and their application to build these precision tracking devices is the second development that makes such studies possible."

What Bigi and Sanda suggested, and what experimenters now do, is to compare the rates for the decay of a neutral B-meson ($B^0$) or its antiparticle ($\overline{B}^0$) to particular sets of particles, for example the set $J/\psi + K_S$. This is yet another version of the car–anticar race. $B^0$ has the quark substructure of a down quark and a bottom antiquark, while $\overline{B}^0$ has that of a down antiquark and a bottom quark. Thus $\overline{B}^0$ would become $B^0$ (and vice versa) if the quarks and antiquarks reversed roles. However, since $J/\psi$ has the quark substructure of a charm quark and a charm antiquark and $K_S$ has very nearly equal amounts of $K^0$ (down plus anti-strange) and $\overline{K}^0$ (antidown plus strange), the final state of a J/psi meson and a K-short meson is a CP mirror image of itself. The experiment of interest compares the rate for the decay B-zero to J/psi and K-short ($B^0 \rightarrow J/\psi + K_S$) with the rate of the decay anti-B-zero to K-short ($\overline{B}^0 \rightarrow J/\psi + K_S$). These two processes are CP mirror images. If these two rates are different in any way, CP symmetry is violated. To remove dependence on details that we cannot calculate well, physicists look at a quantity they call the asymmetry, which is the difference of these two rates divided by their sum. The relationship between the measured asymmetry and the underlying CP-violating parameter of Standard Model theory is simple and clean (which means that the prediction is not subject to any significant corrections from effects that are difficult to calculate).

This decay process has two very special features. The first is that quarks of all three generations are involved: The bottom antiquark (quark), a representative of the third generation, and the down quark (antiquark), a representative of the first generation, are constituents of the neutral $B^0(\overline{B}^0)$ mesons. The charm and strange, second-generation quarks and antiquarks, are constituents in the final state, the J/psi and K-short mesons. The Standard Model has CP violation only in a three- (or more) generation theory (but not if there were only two), so only if all three generations are directly involved can the CP violation be a big effect. Thus these two rates can be very different from each other. Unfortunately this property also has the consequence that the expected rates for this decay are very small; there are many other possible ways for a B-meson to decay that are much more likely to occur. (However, if you can produce enough B-mesons, it is easier to measure a significant difference between two tiny rates than it is to measure a tiny difference between two large rates.)

The second attractive feature of the golden mode is that the theoretical understanding of the decay process is very simple. Once we take the ratio of rates (difference over sum) to form the asymmetry, all the things we do not know how to calculate well simply drop out; they cancel in this ratio. We are left with a very simple prediction for how this asymmetry depends on the Standard Model parameters that appear in the three-generation theory.

## Oddone: How to Build B Factories

It took seven years from the theoretical suggestion until a realistic idea for an experimental facility that could make the measurement of the particle–antiparticle (CP) asymmetry in $B^0 \to J/\psi + K_S$ decay was conceived, and some fifteen years more to design and build these machines and make the measurement.

Indeed, as we said above, when such a measurement was first suggested, it seemed a rather unlikely test. In the 1980s we knew very little about B-mesons. We knew that strange-quark weak decays were suppressed compared to down-quark weak decays, but we did not yet know that bottom-quark weak decays are even further suppressed, a property that makes B-meson decays detectable. Even today this observed fact has no explanation; we can choose the parameters so that the Standard Model follows this pattern, but we do not know why the theory should have this pattern of parameters.

So how do we do this experiment? First, we need to build a *B factory*, which here means any accelerator that produces many millions of collisions that create a b-quark and an anti-b-quark pair. To measure the asymmetry in the decays of a B-meson to J/psi and K-short we must find many events where we see such a decay. Both $J/\psi$ and $K_S$ are easily identified after they themselves decay. For example, the $J/\psi$ can decay to $\mu^+\mu^-$ and the $K_S$ to $\pi^+\pi^-$. These four particles (the two oppositely charged muons and the two oppositely charged pions) can be readily seen and identified in the detector. Once you make enough neutral B-mesons, it is not hard to pick out these decays.

But, since either a $B^0$ meson or its CP mirror the $\overline{B}^0$ meson can decay to this final state, we also need a way to determine which one we had. This we do by looking for cases where, in the same event, the other b-type quark decays in such a way that we can tell whether it was a b-quark or an anti-b-quark. Then we know that the B-meson contained the other possibility, and that labels whether it is a $B^0$ or a $\overline{B}^0$. There are many possible decays that we can use to make this identification from the known patterns of b-quark decays. For example, a b-quark can decay to a c-quark and an electron (or negatively charged muon) plus an (unseen) antineutrino, but the corresponding decay for an anti-b-quark to an anti-c-quark gives a positron (or a positive muon) and a neutrino. So the charge of the electron or positron (or likewise of the muon) produced in such a decay identifies whether the decaying quark was a b-quark or an anti-b-quark. We call such a decay a *tagging* decay. Since we start by making a bottom quark and antiquark pair, tagging one of them tells us the nature of the other. Thus once we have such a tag, it tells us which type of B-meson decayed to the $J/\psi + K_S$ in the same event.

You might notice that we keep saying that such and such a process *can* happen. There are many, many possible sets of particles that can be produced when a B-meson decays. When the b-quark and its antiquark are produced, the thing we are looking for is only one of many possibilities for what happens after that. We have to record many, many events and then sort through them (with computer methods) to find the ones that exhibit the particular patterns we want to study. That is why we need a "factory"; it takes millions of collisions in our accelerator to produce enough events. A reliable measurement of the rate difference patterns for the two possible tag types requires thousands of events of this particular type. To give you an idea of the challenge: For every ten thousand times that we observe a neutral B-meson decay, the desired J/psi K-short state is produced only about three times. The chance that we can also tag the other b-type quark in the same event depends on the type of facility. In the electron-positron B factories, it is about one in three. That too reduces the probability to get an event with the two desired characteristics: the particular B-meson decay we want to study, and a good tag of which type of B-meson we had.

When all the details are said and done, we find that to measure the CP asymmetry to an accuracy of, say, 0.1, an experiment needs to do the following: produce at least five million B-mesons, find the final $J/\psi + K_S$ events, and separate those that come from an initial B-zero from those that come from an initial anti-B-zero. If one does this experiment by colliding electrons with positrons it turns out that one has to measure the asymmetry between the rates as a function of the time between the two B-meson decays. If one simply looks at the total effect, averaging over all time differences, the result is predicted to be zero asymmetry. The difference we want to study is in the time pattern, not in the net effect. This is no easy task.

There are two facilities that have now been carrying out these studies for some years. These are B factories running at Stanford (in California, USA) and at Tsukuba (in Japan). Let us now explore their development and the results they have so far obtained. They will continue to run till about the end of this decade.

Piermaria Oddone, known as Pier, is an intense man, full of energy and enthusiasm for his work (and indeed for life). He was born in Peru, moved to the United States to become an undergraduate at MIT, and got his Ph.D. at Princeton. He has become one of the leaders of the particle physics community. He is now the Director of the Fermi National Accelerator Laboratory (Fermilab). In 1987 he was Director of the Physics Division of Lawrence Berkeley Laboratory and, in addition to his management responsibility, very much an active researcher. He is an expert in both high-energy detectors and in their interface with colliding beam accelerators, having led the development the experimental halls and the program of experiments for the PEP collider at SLAC. PEP (Positron Electron Project) was a storage ring built in the late 1970s at SLAC. Like all such storage rings at that time it was a single ring of vacuum pipe with large magnets around it, designed so that electrons and positrons circulate in opposite directions in the same ring, and thus have the same energy.

In 1987 Pier contributed an important idea. He knew that a colliding ring facility like PEP, then active at Cornell, was producing and detecting the decays of pairs of $B^0$ and $\overline{B}^0$ mesons. However, because these particles were made essentially at rest in the center of the detector, the two B-mesons decayed very close to their production point. There was no way to tell the time between the two decays. Thus the interesting time-dependent asymmetry, even if it was quite large, could not be measured. Pier had the brilliant inspiration that if both B-mesons would be produced moving fast in the laboratory you could make the required time measurement.

In the B factories, electrons and positrons collide at an energy that is just right to produce B-meson pairs at a high rate. By applying simple conservation of energy, one can see that the two B-mesons are the only things produced in these events. Oddone knew this meant you needed different energies (and thus different momenta) for the electron and positron that collide to produce these particles so that they will be moving fast. Electron-positron storage rings, like the one at Cornell, can only circulate equal-energy electrons and positrons in opposite directions around the ring. Pier's inspiration was that it would be possible to do the experiment if you could build two storage rings, say one on top of the other; then you could have countercirculating bunches of particles with different energies. That is the easy part; the tricky part is then to get these bunches, at some point around the rings, to actually pass through one another so collisions can occur, and then steer the remaining particles in the bunches, the ones that did not experience a collision, back into their own rings for another time around and another chance at collision.

If the collision produced a B-meson and an anti-B-meson meson traveling fast enough, close to the speed of light, then the spacing between their decays could be measured, by separating the points from which the tracks of each set of decay products diverge. The fact that the B-mesons are traveling at close to the speed of light in the laboratory effectively stretches out their lifetimes, and thus stretches the typical distances between the decays. If the average distance between the decay of one B-meson and the decay of the other could be made to be a few hundred microns this would allow the two decay points to be separately identified in a modern precision tracking detector. The distance between them would tell the time between the two decays, because we would know how fast they are moving.

What really excited Pier, and many others once he pointed it out to us, was that his back of the envelope calculations said that if you made the electron energy about the energy of the existing PEP ring that he knew so well, and built a second ring for the positrons at lower energy, the measurement would be possible. The PEP tunnel was big enough to fit in this second ring. That was the genesis of PEPII, the accelerator that drives the SLAC B factory. His idea soon excited another group of physicists at the KEK accelerator in Japan. They too had an existing storage ring that could be converted into a B factory. Their project known as KEKB is also now also an active B factory.

The conversion is not simple; it took clever accelerator designers to figure out how to get the bunches of particles from two separate rings to pass through one another, allowing collisions to occur, and then get them back into their own separate rings so that they could cycle around again for another chance at collision. But the suggestion set in motion a scramble to design and build such a facility and its detector. The idea was too good not to be pursued. Both in the United States and in Japan funds were set aside for these projects and multinational teams of scientists began to form to work on them.

Helen was part of the SLAC project from its inception. She says: "It was fascinating to me, as a theoretical physicist, to see how the project evolved from dream to reality. I'd never been this close to an experiment in its early stages before. I was tremendously impressed by the amount and variety of expertise, and cleverness that was needed to turn this idea into reality. Each aspect of the accelerator and each layer of the detector had its own set of experts, or even multiple sets proposing competing solutions to each problem. Choices had to be made. There was international politics, and money, to consider as well as physics in how those choices were defined. The sociology was sometimes as interesting to watch as the physics discussions! Early designs of a detector allowed us to make crude estimates of what could be measured, but all engineering factors such as support structures or space for cables to carry out signals were not included. As the project progressed through the various approval stages toward funding, all these details had to be taken care of and more precise simulations of the detector made to show that it could achieve the desired results."

Both at SLAC and in Japan, the B factories quickly achieved their design performance and have been steadily improving their productivity as they continue to operate. This is a remarkable feat of physics and engineering skill, not to mention dedicated hard work. The storage rings are only part of the story. Once a collision occurs, one needs to find out what happened. So the collision point is surrounded by a large multilayered cylindrical detector of the type described in the preceding chapter. These huge detectors are designed by physicists to provide all the information needed to reconstruct and identify all the particles formed when the B-mesons decay, and to measure their energies and momenta. One needs clever algorithms to decode exactly what happened from all the information about the event obtained by the detector.

In addition there is the challenge of storing and analyzing all the data produced, sorting through millions of collision events to find the needles in this haystack that can tell us about CP violation. The detector, when it is running, is on twenty-four hours a day for weeks at a time. If everything is working well it produces a stream of information containing hundreds of megabytes per second. Fast algorithms must be used to refine this stream, to decide which parts of this are data worth saving into a data base. And then, for each question one wishes to ask, one must access the data and find ways to extract the events that satisfy the particular specifications for the question at hand. Each of the two B factories keeps over 500 physicists around the world quite busy.

The B factories were designed not only to carry out the challenging task of measuring the CP asymmetry in the decay of neutral B-mesons to J/psi K-short states, but also to make similar measurements for a number of other sets of particles that can come from B-meson decays. Many of these sets are produced even more rarely than the $J/\psi + K_S$ case, or are harder to detect. These will require several more years of intense data collection.

One can also make a b-quark and an anti-b-quark by colliding a high-energy proton with an antiproton (or with another proton). In this case the two $b$-type quarks are moving independently and each forms either a B-meson or a $b$-containing baryon. Many other particles are formed in the same collision. These experiments, at high-energy proton-proton colliders can, in fact, make many more B-mesons in a given time than the electron-positron collider B factories. On the other hand, they make many more other things too. The challenge is then to find and study the B-meson decays in an environment where so much else is going on. Only a small fraction of the events can be identified and tagged. The TeVatron at Fermilab can make some interesting measurements on the $B_s$-mesons (particles made from a b-quark and an anti-s-quark or an s-quark and an anti-b-quark). This is another pair of neutral mesons which have interesting properties related to CP symmetries. These particles are not produced at the electron-positron B factories, because they are too massive for the energy available there currently. (Moreover, if one increased the energy enough to make these particles, the rate of collisions would not be high enough to give the number of events one needs to make the studies.) So in many ways the two types of experiment are complementary. Further measurements will be pursued at the TeVatron and at the higher-energy LHC collider under construction at CERN, where plans for a detector optimized for B physics, known as LHCb, are underway. We want to search for discrepancies with the Standard Model theory, and we need to explore the patterns of many different decays to see if they all match that theory or not.

## Running the B Factories: The First Test

The B factories at SLAC and KEK began operation in 1999. At SLAC the experimenters named their massive detector BaBar, after the elephant, because it is large, and because it was to study $B^0$ and $\overline{B}^0$ mesons. (Physicists call $\overline{B}^0$ "B-bar" because of the way it is written; thus, they often refer to $B^0$-$\overline{B}^0$ as "B-B-bar.") At KEK the beautiful (and equally huge) detector acquired the name of Belle. In August 2000, in a conference held in Osaka, Japan, the two experiments announced their first measurements of CP violation in B-meson decays. Both BaBar and Belle had observed a large CP asymmetry in B decays into the particular final state made of the J/psi-meson and the K-short meson. Each experiment separately had enough data to narrow the uncertainty in their result that comes from statistical fluctuation possibilities, and be sure that the effect is real.

New numbers based on larger data samples have been presented as the data are collected. The asymmetry is now known to have the expected dependence on the time between the two decays and a size in the range +0.65 to +0.72. (For the statisticians, that is the plus or minus one standard deviation allowed range.) What does this result mean? How does it reflect on our understanding of CP violation? Have we found any new clues to the puzzle of the missing antimatter?

We can calculate the prediction of the Standard Model for this CP asymmetry. The allowed range depends only on two parameters of the Standard Model theory. The prediction has an uncertainty, but only because it depends on how precisely these parameters have been measured by other experiments. This is what we mean by a clean prediction. We can fix these two parameters in a variety of ways—the amazing part of this story, the success of the Standard Model, is the fact that there are at least five quite distinct measurements, or combinations of measurements, that could be used to fix these two parameters, and they all give roughly the same answer. Furthermore when we put them all together, we get a more precise evaluation of the answer than any subset of measurements can give. When all of the available information is combined, we find that the measured range for the CP-violating asymmetry in the mode $J/\psi + K_S$ overlaps the predicted range. The result is completely consistent with the Standard Model prediction. This was another triumph of the Standard Model, but of course it was a disappointment for those of us who had hoped we might see a clue to new effects by finding a failure of the Standard Model prediction here.

The Standard Model fits the data so far. The Kobayashi–Maskawa mechanism has successfully passed its first experimental test. This means we have as yet no direct evidence for new physics that can make an electroweak baryogenesis successful. Instead we can use the B-meson physics result, in combination with the earlier measurements, to determine the Standard Model parameters more precisely. It also restricts models for new physics because it severely limits any new physics contributions that would change these answers. So is it now worthwhile to continue to look for new physics effects in other possible B-meson decays?

Once you had seen a rabbit in the forest, what would you make of this observation? It would be reasonable to assume that rabbits live in the forest, and it is not just an accidental tourist that you have met. But it would be nice to see a second and a third and perhaps more rabbits to be sure of that (figure 17.1). Indeed, the consistency of the CP asymmetry in neutral B-meson decay into a final $J/\psi + K_S$ state with the Standard Model prediction makes it quite likely that the Kobayashi–Maskawa phase is the dominant source for the observed CP violations. But we would like

Fig. 17.1 It would be nice to see a second, and a third, and perhaps more rabbits.

to see a second and a third and even more measurements of CP asymmetries to state with confidence that we see the Kobayashi–Maskawa mechanism at work. And even when you have seen more rabbits, you cannot conclude that the forest contains no animals but rabbits. Only after you have sat down for a long time, always seen rabbits, and never found animals other than rabbits, is it reasonable to suggest that perhaps they are the only ones present—and even then you cannot be sure.

So it is with our experiments—we must make further tests before we can conclude that it is the Standard Model CP violation and only the Standard Model CP violation that governs CP violation in the world of quarks. The B factories will continue to improve the accuracy of their measurements of all the processes so far studied, and to make new analyses of processes that occur even more rarely than the $J/\psi + K_S$ decay. We have found a first result that agrees with the Standard Model. Further tests are beginning to be possible as more data are collected. As yet these tests are not very precise; they need more data. Studies of B-mesons and their properties will keep physicists occupied for some years to come. We do not know what we will find.

Perhaps in the end we will find exactly the patterns predicted by the Standard Model. That would mean that in all the CP violations that appeared in our experiments, the Kobayashi–Maskawa phase is the only significant player, and no other phases play any significant role. That would make us suspect that the CP-violating interactions that cause baryogenesis are mediated by new particles that are far heavier than the weak force carriers.

There would still a chance that there are new interactions that generate a baryon asymmetry, but that cannot be observed in the B factories, not because the mediating particles are very heavy, but because—like the Z-boson—they do not take part in flavor-changing processes. That answer can be pursued with another type of experiment, the search for electric dipole moments. If we find such effects, effects that would be too small to detect if only Standard Model processes were causing them, then we would have evidence for a new type of CP violation and some chance that baryogenesis related to this CP violation could work.

If neither type of evidence for new CP violation can be found, that would be rather bad news for missing-antimatter detectives like us. It would suggest that the particles that give the CP violation that generates the matter–antimatter asymmetry are extremely massive. If that is the case, we would not be able explore these processes directly in our laboratories—or at least not in the foreseeable future.

We would be much more excited if the results just don't fit together as predicted. Then we would know that CP violation comes not only from the $W^{\pm}$ couplings to quarks, but also from other parts of some extended theory. In that case, the detective work of figuring out just what is going on would swing into action. The data would give us more clues to the mystery of the missing antimatter. Indeed the situation would be even more exciting! If there is CP violation that destroys the Standard Model patterns for the B factory, it would have to be coming from particles that we do not yet know about. So the search for those particles would really be on. They would very likely be produced and studied in other experiments in the not too far future. If we can find them and understand their properties it is possible we can achieve a complete solution of our mystery.

# 18 *SPECULATIONS*

## Why Are We Never Satisfied?

You might think that physicists would be celebrating the successes of the Standard Model. Indeed we do, but we also long to find places where it is wrong and focus much of our efforts on that. Why do that? Partly it is a matter of experience. In the past there have been times when the understanding of physics seemed almost complete, yet the explanation of what seemed a small discrepancy opened up entire new vistas.

Furthermore, there are certain features of the Standard Model that seem to be achieved very artificially. These features can, however, appear naturally, as side effects from larger symmetries. To achieve such larger symmetries, new particles have to be added, and these particles must be massive to explain why we have not yet seen them (or even any indirect effect from them) in experiments so far.

We also know that there are things we have not yet tried to include in this theory. The most obvious such omission is that of gravity. It seems that, to include gravity in our particle physics theories, we have to go not just beyond the Standard Model, but actually beyond quantum field theory. Related to that is the fact that the Standard Model cannot even begin to explain the phenomenon known as dark energy, which causes the expansion of the Universe to be changing in an unexpected way. Further, the Standard Model contains no particles that could be the dark matter that pervades the Universe.

Once we start to speculate on how we might go beyond the Standard Model we work from whatever clues we have. A few of these are empirical information, such as the size of the matter–antimatter asymmetry in the Universe. Others are clues we find in the mathematics, which suggests some possible extensions just from the patterns of what we already know. Such extensions are always speculative. We explore many speculations. But we have measured so much already that many ideas are quickly eliminated because they lead to predictions in conflict with things already known. The theory has to look very much like the Standard model in all the ways where the Standard Model works, and give different predictions only for things that have yet to be checked. That is a serious challenge. Typically it means that the theory is the Standard Model plus some additional things. Additional particles in the extended theory must be quite massive, or interact very, very little, or else they would change things that are very precisely measured.

Physicists have thought of many possible ways to go beyond the Standard Model that keep all the good results. Of course, we do not yet know which, if any, of these ideas gives a correct description of Nature. But it is interesting to explore a little about some of these ideas and their implications for matter and antimatter in the Universe.

## Grand Unified Theories

Grand unified theories are a class of theories that offer a very appealing hypothetical extension of the Standard Model. In these theories, at very high energy, there is only a single type of gauge interactions, that is, interactions where the force carriers are spin-1 particles. The strong, weak and electromagnetic interactions are all aspects of this one unified interaction, differentiated by a spontaneous symmetry breaking. The first model of this type was proposed in the mid-1970s by Howard Georgi and Sheldon Glashow.

Helen remembers: "In 1970, when I first came to Harvard, I had no job and was pregnant. But I had met two of the faculty members, Shelley Glashow and Sydney Coleman, at a summer school in Europe two years earlier, and so I talked them into letting me join their research group without pay. Shelley went on sabbatical leave and I, and soon my daughter, occupied his office for the rest of the year. This was a turning point in my career! The next year I was made a postdoctoral researcher in the group with my own office and a salary, and later I became a Harvard faculty member in that group. Shelley returned from his sabbatical and began to work mostly with Howard Georgi who then had an appointment called Junior Fellow—a high honor for a young researcher at Harvard. He later became a faculty member at Harvard too, and is still there today.

"On some projects I collaborated with both of them. I remember most the fog of cigar smoke when I went to talk to them, particularly if it was just after lunch. (Nowadays I'd complain, back then I just left the room when my stomach got too queasy.) In this period we, and others, were learning how to write consistent theories with gauge symmetries and calculate their implications. A theory for weak and electromagnetic interactions was in place, though it needed the charm quark that had not yet been seen. The theory of strong interactions based on color and gluons was taking shape too. Shelley was full of ideas, every day a new one. Even before the charm quark had been discovered, he was moving on to put the disparate pieces together and see how he could fit them into a larger pattern. I remember him talking about the 'twelve known quarks' (four flavors, three colors of each). Howard was quick to find the flaws in his ideas, and to suggest new variants that might overcome those flaws. Together they made a good team; their work quickly converged to reveal a pattern that is the basic grand unified theory."

To say that there is only a single gauge interaction means that the underlying symmetry of nature is larger than that of the Standard Model. The equations that describe these interactions have a larger set of invariances. They are more restricted, leaving room for only one coupling constant parameter for all gauge interactions. If true, that would make our description of Nature very simple and beautiful. So particle physicists really like the idea of grand unification—it seems almost inevitable.

If the idea of grand unification is realized in nature, the three types of gauge interactions of the Standard Model—strong, weak, and electromagnetic—are just three faces of one and the same force. There is only a single coupling constant and it is universal: if you measure the rate of one process that is mediated by spin-1 force carriers—be they gluons, photons, or W- or Z-bosons—you will be able to predict the rate of any other process mediated by spin-1 force carriers. Because all three are gauge theory interactions it is very tempting to try to unify them in this way.

Why then do we observe three different types of interactions, each with seemingly different strength? As we saw for the weak interactions, we can accommodate a variety of couplings by placing the blame on spontaneous symmetry-breaking effects. The symmetry respected by the ground state of the Universe is smaller than that of the interactions. So a part of the symmetry is hidden. But if we look carefully enough at the features of the three types of interaction, we should be able to see at least some traces of the larger symmetry.

In particular, not any three values for the strong, weak, and electromagnetic couplings would allow us to fit the Standard Model into a grand unified theory. The precise value of the coupling constant depends on the energy at which a specific process takes place. We can calculate how the couplings change with energy. And, if indeed grand unification occurs in nature, the measured values at low energies and the calculated energy-dependent changes of these values should be such that, at some extremely high energy, the three coupling constants become equal to each other.

To have an intuitive picture of this situation, think about trains that always travel with constant velocities and whose routes cross at some junction. If we measure the velocities and location at a certain time, we can say whether they will (or, hopefully, will not) collide at the junction.

For grand unification to be possible, we would like a collision of couplings to occur. Amazingly, the measured values are roughly consistent with this condition. The data in our experiments imply that if we were able to construct accelerators that are $10^{14}$ times more energetic than the ones with which we carry out our present research, all three known types of gauge interactions would be measured to be characterized by the same coupling constant. So maybe the beautiful, theoretical idea of grand unification has really something to do with nature. (This fact was first pointed out by Helen, together with Howard Georgi and Steven Weinberg, in a paper written during that time at Harvard.)

The implications of grand unification go beyond the idea that the three gauge forces are essentially one and the same force. It implies also that the many particle types that constitute a Standard Model generation may be, at root, one and the same thing. Now that remark needs some explanation.

Think, for example, of the down quark. We know that it has a color charge, and that it could come in any of three colors. However, there is no way to tell which of the three colors is carried by a specific down quark that we are observing. This is the consequence of the symmetry: we cannot distinguish a down quark of one color from one of another, because they all react in precisely the same way to all forces, the strong, weak, and electromagnetic. Therefore we think of the down quark as a single particle type, not three different ones; yet they have different color charges, so we could have called them three different particles. In the same way, if the symmetries of grand unified theories were exact, we would not be able to tell, say, a down quark from an electron. They would react in the same way. Only the spontaneous symmetry breaking makes them distinguishable.

But for such a situation to be possible, there are certain constraints that the color, weak, and electromagnetic charges of the particles in each generation have to obey. The clever step of Georgi and Glashow was to see that these constraints are indeed fulfilled and, consequently, there are ways to put the known particles together into grand unified theories. In other words, they succeeded in writing grand unified theories where the predictions, after symmetry breaking, match the observed patterns, giving the quarks and leptons of the Standard Model.

Grand unified theories are a very appealing idea. But to go beyond being an attractive speculation, and become perhaps the new standard, the theory must make predictions that are different from those of the Standard Model and these predictions need to be confirmed experimentally. Indeed, what we said above about the particle unification provides a hint to such a prediction.

If the grand unified symmetry does not distinguish quarks from leptons, there must exist interactions, beyond those of the Standard Model, where a quark can turn into a lepton, and vice versa. This is analogous to the strong interactions which change the color of the quark. When a green down quark emits a gluon, it can become a blue or a red one. The grand unified theory must have, and indeed always has, some additional spin-1 bosons. When a down quark emits one of the new gauge bosons, it can become an electron. In such an interaction, neither baryon number nor lepton number is conserved.

Thus, in contrast to the situation within the Standard Model, grand unified theories predict that an isolated proton is not a stable particle. It could decay to a light meson and a lepton. The precise decay rate depends on the details of the theory. In all of them the rate is extremely slow, some fifty orders of magnitude slower than typical weak interaction decay rates. The reason for that is that the interaction that leads to proton decay is mediated by ultraheavy gauge bosons, heavier by a factor of about $10^{14}$ than the weak force carriers. Thus this interaction is correspondingly ultraweak.

This means that, on average, a proton lives much longer than the present age of the Universe. (This is fortunate for us; otherwise, we would have no protons and neutrons around.) Only very, very rarely does a proton decay faster than that, so the grand unified theories say. The rarity of these events makes an experimental observation of them a very challenging task. But the prospects of observing a proton decay are so intriguing to particle physicists that they took to the challenge and managed to construct experiments that could be sensitive to such rare events. These experiments have observed no signal so far. They tell us that the half-life of a proton is longer than $10^{33}$ years. If it were any less than that, they would have seen some decay.

So there is presently no direct experimental evidence for grand unification. The sensitivity of the experiments is not yet strong enough to rule out the idea altogether, although it has ruled out the first, simplest version of such a theory. The experiments continue, in the hope that a signal will appear, but also aware of the fact that, if it does not, the idea of grand unification will become more and more disfavored. Even more sensitive experiments are on the drawing boards.

What has all of this to do with baryogenesis? The first hint to this question lies in our discussion above. Grand unified theories have do not *oops! pub-lease* respect the symmetry of baryon number. Could the baryon asymmetry be driven by processes that involve the new particles predicted by these theories? Indeed, there exists a simple scenario, called *GUT baryogenesis*, where this is just what happens. (Particle physicists use GUT as an abbreviation for grand unified theory.) In the early Universe, the ultraheavy gauge bosons are produced, along with all other particle types, but they drop out of equilibrium when the temperature goes below their mass. This means there are no longer collision processes energetic enough to produce them, so now the particles of this type gradually decay away.

They decay to quarks and leptons in a way that is baryon-number violating and CP violating, inducing an imbalance between the numbers of quarks and antiquarks, that is, baryon asymmetry.

This was the first model suggested for baryogenesis. When it was first proposed, it was not yet realized that Standard Model processes violate baryon number at high temperature, and that these processes are frequent until much lower temperatures than that for grand unified boson decays. So, at that time, it was believed that there can be no Standard Model baryogenesis. GUT baryogenesis, on the other hand, seemed to work. If true, it would have meant that the baryon asymmetry of the universe is sensitive to physics at unimaginably high energies, well beyond the reach of experiments, and explores cosmological events that took place $10^{-36}$ seconds or so after the Big Bang, when the Universe was unimaginably young.

However, this simple story does not work. There are two reasons for that. First, in most versions of grand unified theories, the new interactions violate both baryon number ($B$) and lepton number ($L$), but in a way that conserves the difference between them ($B - L$). For example, two quarks could exchange one of the GUT spin-1 bosons and become an antiquark and an antilepton. The baryon number changes from +2/3 to −1/3, the lepton number changes from 0 to −1, but ($B - L$) remains a constant +2/3 in this process. Thus, a Universe that started with $B = L = 0$, can evolve through GUT baryogenesis to a situation where $B$ and $L$ are non-zero, but $B - L$ remains zero. In that case, the high-temperature baryon-number-changing processes that occur in the Standard Model play a destructive role: They would gradually erase such a baryon asymmetry and drive both the baryon and the lepton numbers of the Universe back to zero, the value favored by thermal equilibrium.

A second reason that GUT baryogenesis is not the answer to our puzzle is that the idea of cosmological inflation stands against it. In an inflationary scenario, the very early Universe undergoes a period of very fast expansion. In the period of inflation, space expands exponentially rapidly with time, whereas later the growth is much slower. With such rapid expansion, any matter content that had existed before the inflationary period is diluted to vanishingly small densities. Amazingly enough, we now have strong observational evidence, from the patterns of temperature fluctuations in the cosmic microwave background radiation, that such a period of inflation did occur in the very first stages of the development of our Universe. The inflationary period also causes rapid cooling, and it ends with the Universe at too low a temperature for collisions to produce the ultraheavy spin-1 bosons of a GUT theory after inflation is over. Any ultraheavy spin-1 bosons that were there before inflation, or any baryons produced in their decay, would be very rare, diluted away by the huge expansion of space. If GUT baryogenesis were the only source of matter–antimatter imbalance, the present Universe would be essentially empty.

The idea of inflation was in fact motivated by grand unified theories. There is another type of very heavy particle that cannot be avoided in these theories. These are particles that carry a single magnetic pole, known as magnetic monopoles. Such particles have been searched for and are known to be extremely rare. The limit on the density of such particles in the Universe would rule out grand unified theories, if it were not for the early inflationary period that made them dilute. Inflation was invented to get rid of the monopoles. It turned out to solve many other problems of cosmology and gradually became an accepted modification of the Big Bang theory. In the past few years it has received strong observational support from detailed measurements of the cosmic microwave background radiation. It is now the favored scenario for the early Universe. Thus particle physicists and cosmologists agree that GUT baryogenesis is not a likely story for the explanation of the baryon number in the Universe. Grand unified theories, whether right or wrong, cannot solve our mystery.

## Supersymmetry

Grand unified theories are not the only idea that particle physicists have introduced to extend the Standard Model. Another popular idea is to add a symmetry that we call *supersymmetry*. Stated in terms of its consequences, this idea seems very odd; it predicts that, for every known spin-1/2 particle (quark or lepton), there is a related, as yet unseen, partner particle of spin-0. Moreover, for every known spin-1 (force carrier) or spin-0 (Higgs) particle, there is a related, as yet unseen, partner particle of spin-1/2. By related, we mean that the new particle, the superpartner, has all charges—color, electromagnetic, and weak—equal to those of its Standard Model partner. Furthermore, if supersymmetry were exact, these new particles would have exactly the same masses as their partners. But, since we do not see them, supersymmetry must in fact be broken in a way that makes the new particles much more massive than their Standard Model partners.

If this theory is right, so far we know about half of the particle types. This seems like a wild idea, but it is very attractive to theorists. First, it is an interesting additional symmetry that has many attractive properties from a mathematical point of view. Indeed it fits well with even more speculative ideas that go by the name of string theory (it seems that most string theory worlds are supersymmetric). Furthermore, if the partner particles are not too heavy, supersymmetry provides an explanation for certain features of the Standard Model that seem accidental and improbable without this added symmetry. For example, in the context of grand unified theories, it is very strange to have two very different scales for symmetry breaking, one that gives the ultraheavy spin-1 bosons that cause proton decay and another, much lower, scale that gives the W- and Z-boson masses as well as the masses of the quarks and leptons. We call this accident the *hierarchy problem*—what fixes this hierarchy of scales? Supersymmetry provides one possible answer to that question, but it is only really helpful if the new particles predicted by the symmetry are not much more massive than the W- and Z-bosons—perhaps as much as ten times heavier, but not more.

Many particle physicists have studied the detailed predictions of supersymmetric theories, asking what is the best way to find evidence for superpartner particles. Of course, the most convincing evidence will arise if these superpartners are produced in experiments and seen to have the predicted relationship (same charge, different spin) with their low-mass partners. This is one of the goals of the high-energy proton-proton collider LHC (Large Hadron Collider) being built at CERN. The LHC will start its operation in 2008. If the idea of supersymmetry is correct, then it is expected that some of these new particles will indeed be produced there. Further into the future, a high-energy electron-positron collider, the ILC (International Linear Collider) can give a more precise determination of various features of the new particles. In addition to this direct search, there is an ongoing experimental effort to find indirect evidence for supersymmetry. Supersymmetric theories have many additional parameters and can, for example, introduce new flavor-changing and/or CP-violating effects. Such effects are searched for in the B factories, which we described earlier in this book.

You may wonder why we stop with one new idea—can we combine supersymmetry and grand unification into a supersymmetric grand unified theory? Indeed, we can, and perhaps we should: It seems in fact that the merging of the coupling constants works better for such a theory than for one without the superpartners. Perhaps this is a clue that both ideas are right.

One of the most intriguing aspects of the supersymmetric extension of the Standard Model is its implications for the dark matter puzzle. Supersymmetry predicts the existence of particles that can easily explain this puzzle. These are the spin-1/2 partners of the Higgs, photon, and Z-boson, collectively known as *neutralinos*. Neutralinos have neither color charge, nor electromagnetic charge (hence the name neutralino), but they have weak interaction charges. Their masses are expected to be not very different from those of the weak force carriers, that is, about a hundred times the proton mass. These features put them in the category of dark matter candidates that we call WIMPs, which we mentioned earlier. Given their mass scale and their weak interaction features, we know that the neutralinos—if they exist—were in thermal equilibrium when the temperature was high enough. (This fact makes the study of their implications for cosmology independent of the initial conditions.) When the temperature drops below their mass, their density starts to rapidly decrease, because they can still annihilate, but the radiation is no longer energetic enough to pair-produce them. Very soon their small numbers make it hard for them to meet each other and further annihilate, and their number density freezes-out. The theory is predictive enough that we can estimate their relic density—how many of them are around us today—and that fits very nicely what is needed to account for the dark matter. There is much excitement in the particle physics community because, with the combination of the LHC and experiments searching for dark matter, we may learn enough in a few years to understand both how the hierarchy problem of the Standard Model is solved, and how the dark matter problem of cosmology and astrophysics is solved. If that happens, you will surely hear about it because it will bring another revolution to particle physics.

Finally, does supersymmetry change the story of matter and antimatter in the Universe? As mentioned above, supersymmetry has many new sources of CP violation. The supersymmetric partners of the Standard Model particles can induce new and faster processes that distinguish between matter and antimatter. Moreover, supersymmetry can also make the parameters of the bubble right, allowing for a sudden phase transition. Given the large number of new parameters in supersymmetric models, and given that these parameters can assume values in a rather large range, we cannot say for sure whether indeed a sufficiently large baryon asymmetry can be induced. For most of the range of the supersymmetric parameters, the baryon asymmetry is still much too small. We suspect that the solution to the mystery of the missing antimatter lies even beyond supersymmetry. But the nice thing is that we will soon know. If supersymmetric baryogenesis is the right mechanism, some of the new particles must be quite light, and should be observed in the LHC. The LHC is therefore likely not only to shed light on the dark matter puzzle, but also to test the idea of supersymmetric baryogenesis.

## Way beyond the Standard Model

All of the theories we have talked about above fall into the class of theories called field theory, and they do not include gravity in the picture. It turns out that, starting from field theory alone, the task of incorporating gravity is very difficult. No one has figured out a satisfactory way to do it.

That has led to new ideas, collectively known as *string theory*, and, within that genre, the more recent development called *brane-worlds*. These ideas are based on a quite different mathematical structure than field theory, but they give a field theory as their low-energy description. Here the term low energy is relative to the scale set by gravity, which can be an extremely high energy scale, so low energy includes all the things we see with the highest-energy accelerators built to date. However, these theories predict some new effects if you get to a high enough energy to start to probe the true underlying structure of space which, in all such theories, is not just our observed three-dimensional space. There are extra spatial directions in all such theories but, like ultraheavy particles, they are unobservable to us because they curl up on themselves and are ultra-small. The lowest energy that would excite any structure in a small loop dimension is the energy matched to a wavelength that just fits around the loop.

These ideas are quite speculative and very difficult to test. They have introduced a lot of very different pictures for the physics of extreme high energies and thus for the very early Universe too. The rapidly changing multiplicity of ideas makes this a topic beyond the scope of this book. It is possible that these explorations will eventually lead to low-energy field theories that contain quite different extensions of the Standard Model than those suggested from field theory alone. We simply cannot tell yet what the full answer may be. We do know, however, that it cannot be just the Standard Model with one CP-violating parameter and massless neutrinos.

These theories too are constrained by what we do know about the Universe. In particular, while field theory alone can say that the problem of dark energy is simply beyond its scope, any theory that includes gravity must give us a Universe that behaves like ours, with an early period of inflation from which it emerges with the observed dark energy, as well as dark matter and ordinary matter, and its Standard Model interactions, in the right mix. This is no mean challenge, and while there are many speculative versions of these theories that maybe can do just that, there is certainly not one of them yet that is convincing. This is an active and rapidly changing field of research.

# 19

*NEUTRINO SURPRISES*

## Davis, Bahcall, Koshiba: Solar Neutrinos

To return to more mundane extensions of the Standard Model—we already know that neutrinos have mass, and that alone requires us to extend our original Standard Model theory. Indeed, this fact leads us to another, quite different attempt to solve the mystery of the missing antimatter, a scenario that involves neutrinos. It took physics detectives with three different types of expertise to collect the clues that led us to suspect that neutrinos might be to blame for the baryon asymmetry: astrophysicists, particle physicists, and cosmologists. Let us begin with the astrophysics side of the story, and its interplay with particle physics.

Every second, about $10^{12}$ (a million million) neutrinos produced in the Sun cross the palm of your hand. But your palm is transparent to these particles: the force exerted by it on neutrinos is so weak that it is unlikely that even one neutrino from this huge flux stops during your entire lifetime. Not only is your hand transparent to neutrinos; the flux of solar neutrinos goes right through Earth with almost no interruption. Even the Sun itself hardly affects the neutrinos. Most of the solar neutrinos are produced close to the core of the Sun. They stream out with a velocity that is close to the speed of light without being diverted from their trajectories. This unique lack of interaction of the neutrinos, their very elusiveness, is one of the reasons that physicists are fascinated by them—it presents opportunities offered by no other probe.

206

Because neutrinos interact so little, when we observe neutrinos coming from the Sun, we learn directly about the processes that occur in the core of the Sun, whereas we only see light coming from the surface of the Sun. But the same elusiveness is also a source of a huge experimental difficulty: How can we build a detector to trap the neutrinos and measure their properties? The story of these experiments and their surprising results is a great physics detective story.

About six hundred million tons of hydrogen are burned each second to fuel the heat from the Sun. Nuclear physicists have worked for about half a century to understand the details of this process. The burning of the Sun is a multistage process whereby four protons (hydrogen nuclei) get turned into an alpha-particle (helium-4 nucleus), two positrons, two electron-neutrinos, and radiation (energetic photons). These photons are the source of the solar radiation that we experience daily. Heat (infrared photons) and light (visible photons) from the Sun are essential to life on Earth. Much of what we know about the Sun comes from investigating these photons. But because a photon is a carrier of electromagnetic interaction it interacts readily with the charged matter particles in the Sun: the protons and electrons. These change the direction and the energies of the photons by absorbing and reemitting them.

The travel of photons produced near the core of the Sun is so slowed down by these frequent changes of direction that it takes about a hundred thousand years from the moment that radiant energy is produced in a nuclear process until it reaches the surface, from where it proceeds without further interruption to Earth. In contrast, neutrinos that are produced in the center of the Sun make their way out of the Sun with practically no interruption. It takes only about eight minutes from the time that they are produced until they reach Earth, ready to tell us about the burning in the core of the Sun. Moreover, almost all their original features, including their energy spectrum, are preserved.

It would be wonderful to listen to their story too. But to do that, we have to capture them and measure their energy. The experimental challenge is huge. Any measurable event is a consequence of a weak interaction between an incoming neutrino and some target particle, in which the energy of the neutrino is partially or completely transferred, producing moving charged particles that can be detected. Only huge detectors, with multiple tons of target particles, can accumulate enough events to allow a useful investigation. Since only a tiny fraction of the neutrinos interact, even in a very large detector, one needs a very sensitive method to detect these interactions.

In 1964 the experimentalist Ray Davis (1914– ; Nobel Prize 2002) and a young astrophysicist, theorist John Bahcall (1934–2005), proposed an experiment to detect and measure the flux of solar neutrinos. Bahcall had made a detailed model of the processes in the interior of the Sun. He could calculate the rate and the energies at which the neutrinos are created by nuclear processes in the center of the Sun. He also knew that these neutrinos stream out in all directions, and so their rate of arrival, if you could measure it, would provide a probe sensitive to the details of temperature and density deep inside the Sun.

Ray Davis designed the huge experiment, a bold idea for its time. The detector was a tank containing a hundred thousand gallons of dry-cleaning fluid. (The scientific name of this liquid is perchloroethylene, $C_2Cl_4$.) This detector was built deep underground, in the Homestake Gold Mine in Lead, South Dakota. The underground location is crucial for such experiments, to cut down on the rate of events induced by cosmic rays which could mimic the solar neutrino events. When one is looking for a process that happens quite rarely it is important to find ways to get rid of all competing effects.

The idea of the experiment was this: Once in a while, a solar neutrino will hit a chlorine ($^{37}Cl$) atom, and interact weakly with it. This converts the chlorine into argon ($^{37}Ar$) atom. To understand the meaning of "once in a while," realize that the target had about $10^{30}$ atoms of $^{37}Cl$, and only about 10 of them were transformed into $^{37}Ar$ during a whole month of operation. (Bahcall calculated the expected capture rate too.) Periodically the fluid in the tank was circulated through a system designed to find these few argon atoms. One can appreciate the challenge in counting, by chemical means, the few $^{37}Ar$ atoms. This is another extreme example of finding a needle in a haystack.

Davis and his collaborators measured a flux of electron-neutrinos coming from the Sun that is smaller by a factor of about three than the flux that Bahcall, with his *standard solar model*, had predicted. What could be the source of such a discrepancy between theory and experiment? A priori, there are three possibilities. First, it could be that the experiment is flawed. For example, the experimentalists may have overestimated the efficiency of the process by which the neutrinos are captured, or that by which the $^{37}$Ar atoms are found, thereby underestimating the flux of incoming neutrinos that is necessary to induce the observed rate of events. Such an explanation would be the most frustrating but it must be considered. Second, it could be that the astrophysicists have made errors in modeling the processes occurring in the Sun. For example, they might have overestimated the temperature in the core of the Sun, thus predicting fusion rates that are larger than in reality. Such a test of the solar model was precisely what Bahcall and Davis had in mind when proposing the experiment. The third option was scarcely considered in the early days of this story, and that was that there are neutrino properties, if they have mass, that could change the outcome of the experiment. This would be the most intriguing possibility of all.

Since the solar model fitted a lot of other data about the Sun, Bahcall was quite convinced his model could not be changed to fit the neutrino data. John Bahcall gave many talks to convince physicists and astrophysicists that his model was not that adjustable. He also worked hard to challenge it himself, refining and improving the input data about nuclear physics, and proposing other tests sensitive to some of the details of the model. From the usual astrophysics point of view, the agreement between solar neutrino measurements and theory was not so bad. Indeed, the flux is lower than expected, but not by a huge factor. Much of astrophysics is not a precision science; rough agreement means you have a good first model. Thus most particle physicists were skeptical that Bahcall's prediction was precisely correct. For example, the predicted flux of neutrinos is very sensitive to the temperature in the core of the Sun. Davis's experiment was sensitive only to the most energetic of the neutrinos produced in the sun, not to those from the processes that produce most of the radiant energy. If the theory were wrong about the core temperature by only a few percent, it would be wrong about the neutrino flux seen in this experiment by a factor of three. So particle physicists continued to believe that the most likely problem was in the astrophysics model, even though John Bahcall continued to insist that his model was very constrained by a variety of other data inputs and such an adjustment was not possible.

Helen says: "While I cannot speak for others, I can explain my own attitude during these years. The data were a puzzle. After a while it seemed from the cross checks that this experiment was solid, so it needed an explanation. Two possibilities were open but, if I had to bet, I would have bet against the solar model and not against the particle theory. Why? Because the model of the Sun was the kind of complex physics modeling which astrophysicists must do but particle physicists do not. It seemed to me that this model could be wrong in detail even if it was right in its general picture. I could not check any of the details of it myself, and I knew that the flux of neutrinos that Davis could detect was very sensitive to a particular detail—the temperature at the center of the Sun. I knew that another possibility, adding masses for the neutrinos, was always there as an option, but I just was not convinced it was the only option, so I simply said 'I'll wait and see.' My attitude was that astrophysics is not precision physics—I did not know John Bahcall well enough then."

The probability that the source of discrepancy between theory and experiment lay in an experimental error indeed decreased over the years, as further experiments gave similar results. In 1986, a second detector started operating in Japan under the leadership of Masatoshi Koshiba (1926– ; Nobel Prize 2002), the *Kamiokande* experiment. The detector here is very different from Davis's: it is a tank of three thousand tons (one and a half million gallons) of water, located in the Kamioka mine in the Japanese Alps. When a neutrino hits an electron in the water, it can give it high momentum and energy. The fast electron emits radiation which is observed in photon detectors that cover the walls of the tank.

This technique allows not only a measurement of the total rate of neutrinos arriving, the neutrino flux, as in Davis's experiment, but also a measurement of the energy, the direction, and the time of each individual event. Thus, for example, the experimenters could make sure that the neutrinos that they suspected to be of solar origin are indeed coming from the direction of the Sun. The flux measured by the Kamiokande was, like that measured by Davis, significantly smaller than that predicted by the Bahcall standard solar model.

Since then, five more experiments joined the chase after the solar neutrinos: SAGE in Russia, and GALLEX and GNO in Italy, all using gallium atoms as the target particles; Super-Kamiokande, a water detector in Japan, even larger than Kamiokande, its predecessor; and, most recently, the SNO experiment in Canada with heavy water (water with a larger than usual fraction of deuterium rather than hydrogen) in its detector.

In spite of the fact that different targets and various detection techniques were used, all experiments measured a flux of solar electron neutrinos that was smaller than theoretical predictions. Each experiment is sensitive to a different range of neutrino energies, so in combination they probe more than the overall flux; they probe the contributions coming from each of a variety of different nuclear processes in the Sun. The large number of experiments, the variety of detection methods, and the fact that all experiments confirm the deficit in flux make a very convincing case that the explanation of the discrepancy was not in errors by the experimenters.

Likewise, over the years, the possibility that the mismatch lay in details of the solar model grew less and less likely. A big step in this direction was provided by measurements of solar vibrations, or Sun-quakes. The predicted behaviors of these vibrations depend on many of the details of the solar model and are sensitive to the same internal temperature that governs the neutrinos. Solar model experts, such as John Bahcall, became more and more convinced that they could not change the model enough to match the neutrino results without destroying their good fit to other data, such as that on the solar quakes. Furthermore, they pointed out that the rates measured by later experiments were less sensitive to the temperature than the Davis result, because they measured lower-energy neutrinos, from processes that dominate the Sun's overall energy production. Bahcall showed that one could not fit all the multiple measured neutrino rates by simply changing the estimated temperature at the core of the Sun, because different measurements had different sensitivity to that temperature. Bahcall's ongoing attention to the details of his model and to every aspect of the science of this problem showed his strength as a scientist; not content to propose an idea and let others do the work to test it, he stayed involved for over forty years, and again and again contributed crucial refinements to the work. He did the same for more than one other astrophysical problem too.

Yossi says: "I knew John closely from his frequent visits to the Weizmann Institute, where I work. John had warm feelings about this institute since this is where he had met his wife, Neta, when she was a student of nuclear physics (now she is a professor at Princeton University), and he was a visiting research fellow from Caltech. I was immensely impressed with his insights in physics, with the eye that he had for people (many astrophysicists owe their success to John, who identified young talents and gave them opportunities long before anyone else had), his attention to details, and his warmth in personal relations. Our acquaintance grew even closer when I spent a sabbatical year in the Institute for Advanced Study in Princeton, where John was a professor. The weekly "Tuesday astrophysics lunch seminar," bringing together professors, students, and visitors from the Princeton area, initiated and firmly managed by John, was one of the most unusual and exciting scientific programs that I have witnessed. John's authority, stemming from his scientific excellence and commanding personality, was accepted and respected by all. The sound of John ticking with his spoon on a glass caused complete silence among the fifty or so people attending. He gave homework assignments to professors and students alike—read an interesting paper and critically report its essence at the seminar—and everyone obeyed. I learnt more astrophysics by sitting for one year in these lunch seminars than I could have in any other, more ordinary way."

However, the conservative position is not to assume a new result has been demonstrated until the evidence is overwhelming, and so most particle physicists remained on the fence on the question of neutrino masses until dramatic developments in 2001–2002 finally pushed them off. Measurements by the SNO experiment in Canada made it impossible to explain the discrepancy by modifying the solar model. The interpretation of this experiment's results does not depend on any of that model's details. This finally overcame the skeptical conservatism of the particle physicists; now we are convinced that neutrinos do indeed have small masses.

In the SNO experiment, the target contains heavy water, $D_2O$. In heavy water, each of the two hydrogen (H) atoms that make a water molecule, $H_2O$, is replaced by a deuteron (D). While the hydrogen has a proton, the deuteron has a proton and a neutron. This allows two different processes by which solar neutrinos react to be detected in SNO. One is an interaction between the neutrino and the deuteron which produces two protons and an electron. Such a reaction occurs by an exchange of a W-boson between the neutrino and the deuteron: the neutrino becomes an electron and the neutron in the deuteron becomes a proton. Only electron-neutrinos can cause this process. All of the earlier measurements likewise were sensitive only (or at least dominantly) to electron-type neutrinos, which are indeed the type that, nuclear physics tells us, are produced in the center of the sun.

The second process by which solar neutrinos are detected in SNO is an interaction in which the neutrino simply knocks the deuteron apart, producing a moving proton and neutron, and a surviving unseen neutrino. The experiment detects the freely moving proton. This reaction occurs by an exchange of a Z-boson between the neutrino and the deuteron: in this process the neutrino, the proton, and the neutron do not change their identity. This process is equally sensitive to all three neutrino flavors; the electron-neutrino, the muon-neutrino, and the tau-neutrino can each break up a deuteron in this way.

The fusion processes in the Sun produce only electron-type neutrinos. This fact does not depend on the solar model. It is related to the gain in energy from converting four protons into an alpha particle (a helium nucleus—two protons and two neutrons): the mass difference between the initial four protons and the final alpha particle is too small to produce, say, one positron and its matching electron-neutrino plus a muon and the associated muon-neutrino. The energy released in this mass change is only large enough to produce two positrons and the associated electron-neutrinos. If the Standard Model is right that neutrinos have no mass, then all the solar neutrinos arriving at Earth can only be electron-neutrinos. All that an error in the solar model can do is to make a wrong prediction for the flux of these electron-neutrinos. In particular, the flux could have been smaller than the one predicted by the standard solar model. In such a case, the flux measured by both of the two SNO detection processes described above, $W^{\pm}$ exchange and $Z^0$ exchange, would be smaller than expected, and by precisely the same factor: the ratio between the true flux and that calculated in the erroneous model.

On April 20, 2002, the SNO experiment announced the results of the two measurements. The total flux of neutrinos, as deduced using the measurement related to $Z^0$ exchange, is larger by about a factor of three than the flux deduced from the measurement related to $W^{\pm}$ exchange. The experimental accuracy is precise enough to establish that the two measurements yield different fluxes beyond any reasonable doubt. One could not explain any difference between these two measurements by any correction, plausible or implausible, to the solar model. Furthermore, the total flux that this measurement sees is almost exactly that predicted by the solar model for neutrino production (compatible with the model, within the uncertainties of the model and the measurement).

This is a dramatic development: somehow we have the right number of neutrinos arriving, but many of them arrive as the wrong type. Certainly this result can be explained by neither experimental errors nor solar-model errors. The blame is on neither the experimenters nor the astrophysicists. Particle physicists (including Helen) walked out of that talk in 2002 finally convinced. There was only one possibility left: neutrino properties are different from those attributed to them by the Standard Model. Direct experimental evidence that the Standard Model is wrong (or, more precisely, incomplete) had, at last, been achieved. Neutrinos produced in the Sun as electron type arrive at Earth behaving as some other type. By the time of this experiment it was quite well understood that this can happen only if neutrinos have masses. This can lead to the phenomenon observed in these experiments, one neutrino state at production and a different state arriving at a distant detector. We will now explain how this can be so, and what it has to do with the fact that neutrinos must have mass.

## Quantum Neutrino Properties

The neutrino carries neither electric nor color charge. The blindness of neutrinos to the strong and the electromagnetic interactions is the source of their elusiveness. The only significant force that they experience is the weak one. We now must explore the neutrino weak interaction patterns to understand the strange behavior of the solar neutrinos. First let us remember the pattern of quark weak interactions. For quarks we found that the universal strength of the weak interaction was shared or mixed among the quarks of different mass; each up-type quark can convert to any down-type by emitting or absorbing a W-boson, and the interaction rate is shared between these possibilities. If we tried to define the quark state that is produced when a given up-type quark absorbs a W-boson, we would find it is a defined mixture of the three different-mass down-type quarks (down, strange, or bottom).

Within the Standard Model it seemed that the weak interactions of leptons had a different pattern. Each charged lepton converts to a single type of neutrino when it emits or absorbs a W-boson. Indeed, with massless neutrinos, this was the only way to define the three neutrino flavors: we distinguished them by the associated charged leptons when the W-bosons, the charged carriers of the weak interaction, act on them. The neutrino produced when an electron emits or absorbs a W-boson is called the electron-neutrino, $v_e$. The neutrino produced when a muon emits a W-boson is called the muon-neutrino, $v_\mu$. And, similarly, the neutrino produced when a tau-lepton emits a W-boson is called the tau-neutrino, $v_\tau$. We labeled the neutrino type by the type of charged lepton that is involved in the process that produces or absorbs the neutrino and we called this the lepton flavor of the neutrino.

Earlier experiments indeed showed that there are three distinct types of neutrinos, electron type, muon type, and tau type, by observing that neutrinos are always the same type at production and at capture. A neutrino produced in association with a muon could not be captured in a process that produces an electron. All these experiments took place over relatively short distances between neutrino production and neutrino capture. But, as we just saw, somehow this property was not true over the trip from the center of the sun to detectors deep in the earth. How could this be?

Once again we must enter the world of peculiar quantum properties to understand this result. Remember that previously, in our discussion of strangeness, we mentioned that the neutral states that are produced (K-zero and anti-K-zero) always have definite strangeness, but the neutral particles of definite mass (K-short and K-long) are admixtures of two states of different strangeness. This means that the produced state of definite strangeness is a superposition containing both particles of definite mass with given probability. Now we meet a similar situation for neutrinos.

If there are three neutrino particles with three different masses, then the particles that have a definite mass are not necessarily the same as the states that we label by lepton flavor. The electron-neutrino (and each of the other flavors too) corresponds to a particular strength and phase relationship of the three waves for particles of definite mass. The first to realize that this leads to the possibility of an effect such as that seen for solar neutrinos was Bruno Pontecorvo (1913–1993), in 1957. Neutrinos, if traveling far enough, can undergo a quantum effect known as flavor oscillation.

The thing that makes waves very different from our classical notion of particle behavior is the phenomenon called interference. If two waves arrive in phase with one another, they can add to give a much bigger wave but, if they arrive out of phase, they can cancel one another out. The phenomenon of particle interference effects is totally counterintuitive if we think about classical particles, but by now it has been clearly documented in many beautiful experiments. One more thing that you need to know is that, as a particle travels through space, the quantum property that is equivalent to the phase of the wave changes in a way that depends on the particle mass and energy—these are what determine the wavelength of the wave.

This means that a neutrino can start out as a superposition of mass states with the waves adding up in such a way that it is exactly the right combination to be an electron-neutrino, and canceling any option for a muon- or tau-neutrino. That is what we mean when we say we produce an electron-type neutrino. But, as this state travels, the phases of waves for the various mass states change differently; the three different mass components of the state have ever so slightly different wavelengths. This changes the way the waves combine, so that the probability to find a tau- or a muon-type neutrino can increase, while the probability to find an electron-type neutrino decreases, and somewhere along the path it can even vanish. At that point the electron wave is canceled between the three mass states. (Because the wavelength difference of the waves corresponding to different mass neutrinos is tiny, it takes a very great distance for this to happen.) Twice as far along, the waves get back into their original phase relationship, and once again you see only electron-type neutrinos. This is why we call it a flavor oscillation, rather than just a flavor change.

When neutrinos propagate in dense matter, as they do in the interior of the Sun, then there is an additional twist on the phenomenon of neutrino flavor oscillations. The final result is the same, a conversion of one neutrino flavor to another, but one must consider interactions with the matter as well as the phase mismatches that arise from mass differences. This second effect was first understood by two Russian physicists, S. Mikheyev and Alexei Smirnov and, independently, by the U.S. physicist Lincoln Wolfenstein; hence it is known as the *MSW effect*. For the neutrinos arriving at the Earth from the center of the Sun, both the MSW effect in the Sun and the flavor oscillation that occurs as they travel from the Sun to the Earth must be taken into account to find what flux of each type of neutrino we should expect at the Earth for a given pattern of neutrino states of definite mass.

Let us recall some facts about solar neutrino experiments. First, most of the solar neutrino detectors are sensitive to only $v_e$'s. If $v_\mu$'s or $v_\tau$'s hit the detector of any of the SAGE, GALLEX, GNO and Homestake experiments, they leave no signal. Similarly, the $W^\pm$-exchange processes in SNO can only detect an electron-neutrino. The only exceptions are the water detectors Kamiokande and Super-Kamiokande, with high sensitivity to $v_e$ and small sensitivity to $v_\mu$ and $v_\tau$, and finally the $Z^0$-exchange process in SNO which is equally sensitive to all three flavors. Also remember that the nuclear burning of the Sun produces only electron-neutrinos.

Combining all the pieces of the puzzle together, we find there is only one simple explanation for all experimental results. The burning of the Sun indeed produces a huge flux of electron-neutrinos, as predicted by Bahcall's standard solar model. However, on their way from the production point within the Sun to the detection point on Earth, the solar neutrinos oscillate and some of them change their flavor. The total flux of solar neutrinos that arrives at the Earth fits the theoretical prediction, but part of it arrives as muon- or tau-neutrinos.

Additional (and no less convincing) evidence for neutrino masses comes from a quite different set of experiments, measuring the flux of atmospheric neutrinos. These experiments were also carried out by the Kamiokande detector. Atmospheric neutrinos are produced by the collision between cosmic ray particles (primarily protons, coming from outer space) with particles in our atmosphere. By modeling these processes, one can calculate the expected mix of electron-type and muon-type neutrinos that are produced by particle decays after these collisions. The experiments find that the neutrinos that arrive at the detector from above have the expected mix of muon- and electron-type neutrinos. But when studying neutrinos that arrive from below, having traveled all the way through the Earth, one sees fewer muon-type neutrinos than expected. This too can be explained by neutrino oscillation, in this case of a muon-type neutrino to a tau-type one (which is not detected). This gives additional and different information about the neutrino states of definite mass than the solar neutrino data. By putting the information from both experiments together, we can determine that there are indeed three distinct mass states and measure the mass-squared splitting between them. What we do not yet know is the absolute mass scale, but the mass-squared differences give us a minimum value for each of the two heavier neutrino masses, and that is definitely not zero.

The conclusion is that neutrinos have masses, and the neutrinos with well-defined masses are quantum mixtures of the neutrinos with well-defined weak interaction. The oscillation experiments tell us the values of the mass(-squared) differences. Indeed, since we have now seen flavor change, not only for electron-neutrinos made in the Sun but also muon-neutrinos made by cosmic ray processes in the upper atmosphere, we have measured two different mass differences, and thus know that at least two of the three masses are different from zero—and that suggests that most probably all of them are.

Until these two sets of results (for solar and atmospheric neutrinos) became clear, all experimental evidence allowed the possibility that neutrino masses were zero. Experiments searching for more direct signals of neutrino masses have so far achieved only upper bounds on the value of such masses. These bounds are good enough to show that neutrinos masses must be much smaller than those of other matter particles. We conclude, from all the evidence we have to date, that the neutrino mass may be smaller by a factor of about a hundred million from the mass of the lightest charged particle, the electron. Indeed—this is a very tiny mass, but its implications are far reaching.

The Standard Model predicted that neutrinos are massless. The puzzle of the missing antimatter was one important reason to suspect that this Standard Model is an incomplete description of nature. The existence of neutrino masses is now another. There are a number of other reasons to think that the Standard Model is not the full theory. To explain various puzzles that are unanswered within the Standard Model, physicists invent various models, extensions of the Standard Model, in which such problems are solved. We described some of these ideas, and why they are attractive, in the previous chapter. But it is not enough to like a model for theoretical reasons. To decide that a model is relevant to nature, experimental evidence must indicate a need for it too. An example of such evidence would be the discovery of a particle that is not part of the Standard Model. Of course, measuring a property of a particle that is inconsistent with the prediction of the Standard Model also gives evidence that this theory must be extended or changed in some way, and can, like a newly discovered particle, point the way to the next, more complete theory.

So what does neutrino mass tell us? Interestingly, almost any extension of the Standard Model into a viable grand unified theory gives neutrinos nonvanishing masses. The relevant ingredient in these extensions, which does not appear in the Standard Model, is that they all include elementary fermions (that is, particles with spin-1/2, like neutrinos) that are insensitive not only to the interaction of the strong and electromagnetic forces but also to the weak force. Almost all grand unified theories predict such particles. If such a particle exists, a particle that does not interact via the Standard Model gauge interactions (and is therefore often called sterile), it induces a mass to the Standard Model neutrinos because the Higgs particle has Yukawa-type couplings that connect Standard Model neutrinos to these new heavy sterile particles.

Furthermore, if the new, sterile particles are very heavy, as predicted by most of the extensions of the Standard Model, then the induced mass of the Standard Model neutrinos comes out to be much lighter than that of any charged particle. The heavier the new particles, the lighter the neutrinos would be. The idea that the Standard Model neutrinos gain very small masses through Higgs interaction with very heavy particles became known as the *seesaw mechanism* (one mass goes up when the other goes down).

Under such circumstances, where the Standard Model predicts that neutrinos are massless, but most of its extensions predict, via the seesaw mechanism, that neutrinos are very light but definitely not massless, the experimental search for neutrino masses carries huge importance. Experimental results require that neutrino masses are not all zero, and the Standard Model does not provide a full description of nature. suggests new elementary particles beyond those of the Standard ve can measure the actual value of the mass of the neutrinos, ch us something about the scale of masses of the new particles. nmarize three facts about the properties of the neutrinos that icists are able to conclude when they examine the combined solar and atmospheric neutrino experiments. First, neutri- es. Moreover, the three neutrino masses are different from ond, neutrinos mix. Each of the three massive neutrinos n of the three neutrino flavors or, to say it another way, each of the charged leptons produces a mix of neutrino mass types when it absorbs a W-boson. This is the same type of pattern as we saw for quarks, though the details of the mixing pattern are quite different. Third, the masses of the Standard Model neutrinos are significantly smaller than the masses of any charged fermions. Very likely, the heaviest of the neutrinos has a mass that is about a million times lighter than the mass of the lightest of the charged fermions (the electron).

From these three facts, theorists can use their theories to build further, plausible, deductions. Here are three such *plausible* predictions (not yet experimental facts). First, we predict that the laws of nature for leptons and antileptons are different. This would certainly not be possible for the Standard Model interactions that involve leptons, had neutrinos been massless. But given that they have masses and that they mix, there is no reason that the weak interactions would not be different for leptons and antileptons in much the same way as they are different for quarks and antiquarks. It is a challenge for future neutrino experiments to check whether this is so.

The next two predictions arise if we assume that the neutrinos are light because of the seesaw mechanism, the mechanism that appears in almost any extension of the Standard Model and gives the tiny masses to the light neutrinos. Indeed, this pattern in our speculative theories was the reason that many physicists suspected that neutrinos might have masses even before there was convincing evidence for that. So now, that there is such evidence, it is reasonable that we take it to support the seesaw idea. But when we do so, we are led to make two more predictions. The second prediction says that heavy sterile fermions exist. By sterile we here mean that these heavy fermions are insensitive to any of the strong, electromagnetic, and weak interactions; they interact with the known particles only via gravity and Higgs effects. The existence of such fermions is crucial for the seesaw to work; it is their heavy mass that drives the masses of the active neutrinos (that is, those that are sensitive to the weak interactions) to be so light.

The third prediction is that lepton number is not conserved in nature. This last prediction follows from the patterns of the seesaw mechanism itself. The seesaw mechanism requires that, on one hand, the heavy fermions have heavy masses and that, on the other, Higgs particle Yukawa couplings connect the light neutrinos to these heavy ones. If we try to define the lepton number of our new heavy fermions we find a contradiction. If we choose to say that the heavy fermions carry zero lepton number, then their masses will respect lepton number conservation, but their Yukawa interactions will not. Or, one could say that the heavy fermions carry one unit of lepton number; now the Yukawa interactions respect lepton number, but the heavy masses break the symmetry. It is impossible to assign a lepton number to the new particles that would be consistent with lepton number conservation for all terms in the field theory energy function. Thus we know that lepton number is not a conserved quantity in a theory with such particles.

Physicists around the world are working hard to design experiments that could test these ideas with the requisite (exquisite) sensitivity. The heavy fermion of the second prediction is so heavy, and interacts so little, that it is very unlikely we can ever see it directly (but people once thought neutrinos would never be detected either). We can, however, look for CP violation or for lepton number nonconservation which come about because these heavy particles contribute in quantum intermediate stages of some process. These will be tiny effects, very rare processes. But any sign of such a process will be clear evidence that these ideas are on the right track, and that some such heavy particles exist.

The exciting point is that the new facts that we have learned about neutrinos, and the predictions that follow from these, lead to a totally different possible answer to the mystery of matter–antimatter asymmetry in the Universe.

# 20 FOLLOWING THE NEW CLUES

Could the asymmetry between matter and antimatter in the Universe, observed through its baryon content, have its origins in an asymmetry in the lepton content of the universe? The answer is—maybe. This possibility, first suggested by two Japanese physicists, Masataka Fukugita and Tsutomu Yanagida in 1986, has been named *leptogenesis*. At present it is the prime suspect for solving our matter–antimatter mystery, the most plausible scenario to explain the puzzle of the missing antimatter. How does it work? Why did it become an attractive scenario only recently? What can we learn (and, no less important, what can we not learn) in the future that would close in on leptogenesis as a solution to the puzzle? These questions are the subject of this chapter.

## Some Things We Know

Even in the Standard Model, baryon number is not an absolutely conserved quantity. The theory predicts baryon-number-changing processes that are so extremely rare at low temperatures that we can say they are effectively absent. However, these processes are as likely as ordinary (baryon-number-conserving) processes when temperatures and particle densities are high enough, as they are in the early Universe. These same Standard Model processes also change lepton number.

While this peculiar high-temperature Standard Model process violates the conservation of both baryon number ($B$) and lepton number ($L$), it does so in a way that respects the conservation of baryon minus lepton number ($B - L$). In other words, the number of baryons plus antileptons minus the number of antibaryons plus leptons is never changed by any Standard Model process. Unlike most of the conservation laws that we know, this one seems accidental; it has no special symmetry related to it, but just happens to be the case once all the pieces of the Standard Model, and all its other symmetry properties, are put together. (This is very much like the CP symmetry that would accidentally hold in the Standard Model if it had only two generations.) Physicists find this an intriguing situation; it causes them to wonder whether this property is indeed just accidental, or whether it is inherited from some symmetry in a larger theory for which the Standard Model is just a good approximation at the low energies at which we can conduct our experiments.

Let us consider a specific example of how $B - L$ conservation might work out. Imagine that we start from an initial state that has some lepton number that is not zero, but zero baryon number. For the sake of concreteness, take the initial state to consist of three neutrinos, one of each flavor type, $\nu_e$, $\nu_\mu$, and $\nu_\tau$. If the three neutrinos collide, the collision could yield a very different final state. One possibility, so the Standard Model says, is a final nine-antiquark state: anti-up, two anti-down-, anti-charm-, two anti-strange-, anti-top-, and two anti-bottom-quarks. As promised, the baryon number minus the lepton number has not changed: it is $-3$ in both the initial and final states (remembering that each antiquark carries a baryon number of $-1/3$). But, while initially we had a net lepton number and no net baryon number, the final state of this process has net (negative) baryon number and no net lepton number. From what you know about the rarity of neutrino interactions, this process seems absurd: how often will three neutrinos collide and interact? In any experiment we could mount today the answer is literally never. But it turns out that such a process, and others like it involving multiple particle weak interactions, are not so improbable, if the conditions are different from the ones that characterize our experiments. In particular, such processed cannot be disregarded in the hot, dense early Universe.

Indeed, the Standard Model tells us that processes of this type occurred when the Universe was young enough. They open up an entirely new logical possibility. It could be that some processes from physics outside

Fig. 20.1 We, the physics detectives, have been searching for suspects in the wrong place.

the Standard Model induced a net (negative) lepton number and, later, the Standard Model baryon- and lepton-number-changing processes transformed some of these antileptons to produce a net positive baryon number, thereby giving us the observed baryon-antibaryon asymmetry in the Universe. Given any definite initial lepton asymmetry, one can actually calculate what the final baryon asymmetry would be.

What remains to be done is to check whether plausible scenarios for physics beyond the Standard Model can produce a lepton asymmetry of the right size to give the observed baryon asymmetry in this way. Below we give an example of how this might have happened. One needs physics from beyond the Standard Model for this initial step, because within the Standard Model no such process occurs.

The whole scenario reads like a good detective story. We, the physics detectives, have been searching for suspects in the wrong place (figure 20.1). We were focusing on the physics of quarks, while it could be the leptons that are to blame for the survival of matter in the Universe.

## Some Things We Speculate About

The new understanding of neutrino physics that has been achieved in the last few years is crucial for our march along the road to a theory that allows leptogenesis. We now present a simple set of ideas for expanding on the Standard Model that could lead to successful leptogenesis. By "successful" we mean that the new model, with reasonable assumptions about the values of its parameters, can produce a lepton asymmetry in the very early Universe that can lead to the observed baryon asymmetry. The test of whether this story is true or not then comes from tests of other predictions that this model makes, predictions as to how physics is different from that predicted by the Standard Model.

Imagine that we add a new species over and above the Standard Model list of particles. This new species is different from all the fermions that we know to exist because we assume that it is not sensitive to any of the three types of Standard Model gauge interactions: It does not carry an electromagnetic charge, so it does not interact electromagnetically; it carries no color charge, so it does not respond to the strong interactions; and, finally, unlike the known neutrinos, it does not even have a weak charge, so it does not couple to the weak force carriers, the W- and Z-bosons. The fact that it has none of these interactions makes it obvious why we have never seen such a particle—how could we look for it? What is less obvious is the fact that such particles could be produced in large numbers in the early Universe, and hence could play an important role for the history of the Universe.

There are three things that such a hypothetical particle can do that are relevant for leptogenesis. First, it could be very heavy, with a mass that is orders of magnitude larger than that of all known particles. In the expanding Universe, such particles, because they interact so little, go out of equilibrium with all the other particles and radiation as soon as the temperature of the Universe goes below their mass. (If talking of temperature in terms of mass seems strange, notice that we can convert from temperature to energy using Boltzmann's constant $k$, and from mass to energy using the speed of light $c$ and Einstein's mass-energy relationship.) What this means is that, at that temperature, the typical collisions of lighter particles are not energetic enough to produce the heavy particles. After that whatever population of them is present evolves only by the expansion of the Universe and by their decays. For the particles to play a role in leptogenesis, they need to be out of thermal equilibrium and then to decay very early on, after inflation but well before the baryon- and lepton-number-violating processes of the Standard Model become too slow to be effective in converting some of the lepton asymmetry to a baryon asymmetry. A very large mass allows for such an early decay, so it is a desirable property here. (Note also that, since they decay, none of these particles are around today. So, despite the fact that they are massive and interact very little, two properties we told you were needed for dark matter, these particles cannot be the explanation for dark matter. Dark matter must be made of some more stable type of particle.)

Second, while our particle has none of the Standard Model gauge interactions, it can still have Yukawa interactions. A particle with neither electromagnetic charge, nor color charge, nor weak charge, can couple to the ordinary neutrinos and the Higgs particles via such an interaction. Since our hypothetical particle has no other interaction, it can decay only in one way: to a light lepton and a Higgs particle. It is important here that, since there are three light neutrinos, these Yukawa couplings can violate CP symmetry. Thus, the decay rate of the heavy particle into a final state that includes a neutrino could be different from the decay rate into a final state that includes an antineutrino.

Third, lepton number is not conserved in any theory with such a particle. To conserve lepton number we would have to assign the new particle a lepton number such that its Yukawa interactions would be lepton-number conserving. This might be expected, in analogy to the fact that the Yukawa interactions of charged lepton conserve lepton number and those of quarks conserve baryon number. Thus we would like to assign lepton number +1 to a heavy particle that decays into a neutrino (plus a Higgs particle) and lepton number −1 to its antiparticle, that is, the heavy particle that decays into an antineutrino (plus the Higgs antiparticle). We find, however, that when we write into the energy function the terms that can give a large mass to the new particle, these terms are of a new type. They mix the new lepton (as defined above) with the antilepton. This means that the heavy state with a well-defined mass does not have a well-defined lepton number—here we have a case where the totally neutral spin-1/2 particle (neutral to all gauge interactions) and its antiparticle are really not distinct objects. (To introduce another name, these are called *Majorana fermions* after Ettore Majorana (1906–1938), the Italian physicist who pointed out this possibility shortly after the Dirac equation, with its distinct fermion and antifermion, had been understood.) In terms of physical processes, one and the same heavy state can decay either to a final neutrino (plus a Higgs particle) or to a final antineutrino (plus a Higgs particle). These two possible final states have different lepton numbers. There is no way we can assign a lepton number to such a particle so that its decays do not violate conservation of lepton number.

But now we recall the three Sakharov conditions for baryogenesis, or rather their analog for leptogenesis: lepton number violation, C and CP violation, and departure from thermal equilibrium. The new particle just begs to satisfy all three conditions. If it has the right mass and the right lifetime, which is easy to arrange in reasonable models, it will drop out of equilibrium. And then it will decay in a way that violates lepton number, C, and CP, thus inducing a lepton asymmetry.

The crucial point is that, given lepton number violation and CP violation, when the new-type particles decay, more of them decay into antineutrinos than into neutrinos. A lepton asymmetry will be induced, with negative lepton number. A little later the Standard Model interactions take over, and processes that change both baryon number and lepton number but not their difference convert the lepton number into a combination of lepton number and baryon number, with a positive baryon number that can be of the right size. Once the Universe cools sufficiently, the baryon- and lepton-number-changing processes become extremely rare, and we arrive at the situation we see today: A Universe containing matter, but very little antimatter, in which protons are effectively stable and thus matter persists.

## Fitting It All Together

So here is a list of things that we threw into our hypothetical theory of nature in order to make leptogenesis happen: A heavy particle, with no strong, weak, or electromagnetic interactions, a mass that violates lepton number, Yukawa interactions with the light neutrinos, and CP violation in the Yukawa interactions. "But wait a minute," says the alert reader, "I have seen this list of things before!"—and right she is!

The same list of things appeared before. But it appeared not as a wishing list for leptogenesis. Earlier it was the list of plausible predictions that follow from the observations of atmospheric and solar neutrinos. When a plausible prediction, based on new experimental data, provides also a possible solution to a long-standing puzzle, we think that we may be on the right track. For particle physicists, this is a reason for excitement. Of course, so far the evidence for leptogenesis is, at best, circumstantial. Like good detectives, we would like to do better than that. Leptogenesis will remain forever a hypothesis if we cannot observe some of the relevant processes directly in laboratory experiments. Can we?

The short answer is *no*. There is a limit to how high in energy the collisions can be in accelerators that one can imagine building (certainly in our lifetimes, and possibly forever). The hypothetical particles that induce leptogenesis in their decays are many orders of magnitude heavier than any foreseeable accelerator can produce. This difficulty is even more extreme, since none of the known gauge interactions will produce the new heavy leptons. If we cannot produce these particles, we cannot watch them decay, we cannot measure the decay rate, and we cannot tell whether indeed the CP violation in these decays is of the right size to produce a lepton asymmetry that is neither too large nor too small to later induce the observed baryon asymmetry. In short, the scene of the crime appeared once in the history of the Universe and it is hopeless to think that we can reconstruct it in our laboratories.

But there is also a longer answer. While it remains true that no direct tests of the leptogenesis idea are possible, it is still possible to get more, and better, circumstantial evidence than we have today from future experiments. In particular, two of the ingredients that are crucial for this story to be true, lepton number violation and CP violation in lepton interactions, are, at present, not experimental facts but just plausible consequences of what we know. First, neutrinos are massive and mix, and, therefore, there is no reason to expect that CP is respected by their interactions. Second, they are very light, which, in all likelihood, can be related to the existence of a heavy neutral fermion that introduces lepton number violation. Experimental evidence of either or both of these ingredients, namely, lepton number violation and/or CP violation in neutrino interactions, would be important support for the story we have just concocted.

One process that may one day exhibit lepton number violation, and possibly also allow us to measure the actual values of the three neutrino masses, is a rare effect called *neutrinoless double beta decay*. Remember that beta decay is a standard weak decay process of a nucleus. It involves the decay of a proton to a neutron, a positron, and a neutrino. Double beta decay is a less common but experimentally observed process that involves the transition of two protons into two neutrons, two positrons, and two neutrinos. Neutrinoless double beta decay is a third possibility for nucleus decay that has been searched for but, so far, never observed. Here two protons transform into two neutrons, two positrons, and nothing else (in particular, no neutrinos). Such a process is consistent with all the gauge symmetries of the Standard Model. For example, electromagnetic charge is conserved because the electric charge of two protons is the same as that of two positrons, while the neutrons are neutral. But this process violates lepton number. If, indeed, lepton number is violated by neutrino masses, the process should occur. Otherwise, it would not.

Given the lightness of the active neutrinos, the process is expected to be very rare. But even the light neutrino mass may be just enough to induce neutrinoless double beta decay at a level that would be observable in future experiments. If that happens, then lepton number violation will become an experimental fact and there will be little doubt that the seesaw mechanism for giving neutrinos a tiny mass by adding a very heavy neutral (Majorana-type) fermion to the theory is indeed realized in nature.

CP violation in the lepton sector will also be searched for in future experiments. One can use accelerators to create neutrino and antineutrino beams. By a multistep process one can make a beam that is close to a pure electron-neutrino beam, or alternatively one that is purely anti-electron-neutrinos. Detectors located at very large distances from the source can be designed to be sensitive to both anti-muon-neutrinos and muon-neutrinos. These produce anti-muons and muons, respectively, in the detector when they interact with nucleons in the target volume. Thus one can try to measure the rate at which electron-antineutrinos oscillate into muon-antineutrinos and the rate at which electron-neutrinos oscillate into muon-neutrinos. If these two rates are measured to be different from each other, CP violation in leptons will be discovered. This experiment will not be easy. Whether it can be achieved with accelerators now being built or not depends on one detail that we do not yet know: Precisely how much of an electron-neutrino is there in the heaviest of the three light neutrinos? If this component is too small, the violation of CP will be too tiny to make an observable difference between the rates. So our next task is a set of experiments to pin down that detail, to measure the small fraction of electron-type neutrino in a definite mass state that we know is predominantly muon and tau type.

The puzzle of the missing antimatter has more than one possible answer. The leptogenesis answer is a newly attractive possibility. It says, quite remarkably, that there are considerable amounts of matter in the Universe because the laws of nature are different for leptons and antileptons. The surprising results from observations of solar and atmospheric neutrinos make this scenario plausible. Before we had strong evidence for neutrino masses, most physicists regarded this scenario as just too speculative. But now the evidence for neutrino masses and mixing takes us in this direction, while the evidence from the B factories continues to give just Standard Model physics, making it harder and harder to concoct an extension of the theory that fits those data and gives the asymmetry via baryogenesis. Together the two developments have swung the pendulum of our hunches toward the leptogenesis idea. But we do not yet know. Both baryogenesis and leptogenesis require physics beyond the Standard Model, and indeed beyond the ability of experiments up to now to discern it. Future experiments can support or constrain each of these ideas, these hypothetical additions to the theory of particles, perhaps even constrain them so strongly that we are forced to give them up.

In the future, the plausibility of leptogenesis could increase. In particular, we may observe the laws of nature that distinguish lepton matter from lepton antimatter in action. But there is a limit on what we can prove here. We will probably never be able to observe all the processes that contribute to leptogenesis directly in our laboratories. Leptogenesis may be an attractive and reasonable solution of our puzzle, but it seems likely that it will remain forever a hypothetical solution, or else be ruled out by other evidence. It will be very hard to confirm that this idea is correct.

# 21 *FINALE*

So we come to the end of our book, but not to the end of the tale. We have explored some grand themes: the Universe; matter and antimatter; energy; symmetry and universality. These themes are interwoven together in the fabric of physics, developed and explored by experiments in our particle laboratories and in space. Anything we learn about one affects our understanding of the others. Of course we must rely on these experiments and observations to test and refine out theories, guided by whether our ideas can describe the world around us. Carefully designed probes must be made, alert for contradictions between our best current theory and the evidence. Every detail counts! Any good detective needs a theory, but only one alert for contradictory strands of evidence will ever come up with a theory that allows the mystery to be solved.

As yet this mystery remains unsolved. Cosmologists speculate about the history of the Universe, their theories constrained and informed by a growing body of observations directly probing the early Universe, as well as by the established theories of particle physics. Particle physicists speculate about extensions of the Standard Model, their theories constrained and informed, not only by the growing body of results from particle experiments but also by the applications of those theories in astrophysical and cosmological contexts.

Three big questions stand at the edge of our current understanding, waiting for new ideas or new experiments that will clarify them. What is dark energy? What is dark matter? How and when did the imbalance between matter and antimatter occur? In this book we have explored them all a little, though our emphasis has been on the last of these three. In both particle physics and cosmology, and especially in those areas where they meet in explaining the early Universe, there has been an era of great progress. But these unanswered questions remain; questions honed and refined by the progress already made, but questions still, mysteries that have not yet yielded to our investigative skills. We can offer scenarios but we cannot as yet give answers.

It has long been so, and perhaps will long continue to be so. What changes are the questions. Each time we answer some of them we gain the language in which to formulate new questions, questions that earlier could not even be stated. Prior to Dirac there was no concept of antimatter, and so no mystery about its absence in a matter-dominated world. But progress in understanding matter led inevitably to the knowledge of antimatter and revealed this deep mystery. Who knows what new mysteries its answer will unveil? We can only dream.

What we do know is that, in asking these questions, questions prompted by the universal human impulse to wonder, we have learned much. We are convinced that further experiments and further probes sensitive to distant objects and conditions in the Universe will continue to teach us more. The interplay between the very smallest things that exist, studied in high-energy accelerator laboratories, and the very largest reaches of space and time in the Universe will illuminate both disciplines and continue to teach us things more "rich and strange" than we could invent in our wildest imaginations. We are privileged to spend our lives in this fascinating pursuit. We hope you have enjoyed the glimpses of it that this book provides.

## Perspective

Much of the particle physics part of this chronology was originally written by Max Dresden, a friend, physicist, and historian of physics, who knew many of the great men of the early days of quantum physics and gave many delightful lectures about them to high-school teachers in a program that Helen ran at Stanford when Max was a long-term visitor there during his retirement. Part of this chronology was also published in the book *The Charm of Strange Quarks*, by R. Michael Barnett, Henry Mühry, and Helen R. Quinn, published by AIP Press.

Max said: "This appendix presents a historical perspective which shows how the continual interplay of theory and experiment, and the development of new technologies, led, after many unexpected twists and turns, to the theory of the basic structure of matter now called the Standard Model. The Standard Model can legitimately be considered as the culmination of the investigations on the structure of matter that started in the 1800s. Of course the ideas at that time have their own historical roots, stretching all the way back to Democritus, who discussed the idea of indivisible (fundamental) particles which he called atoms in 400 B.C. The chronological summary collected here is intended to give an idea of how,

233

in spite of misunderstandings and mistakes along the way, the present scheme evolved."

The last 35 years have been a period of testing and consolidation for particle physics, as physicists on the one hand learned how to calculate predictions from the Standard Model theory, and on the other built new or improved accelerators to test the predictions in many ways and with increasing precision. Time and again the results have matched the predictions. Physicists today are in some ways frustrated by how well the theory works, because we will get a glimpse of something new only when current theory fails to predict the data. The one really new element for particle physics in recent years is the recognition that neutrinos are not massless particles. This requires an extension of the Standard Model theory.

The cosmology items in this timeline are a new addition, compiled chiefly using web searches to augment our own limited knowledge of this history. One particularly useful site was the "Brainyencyclopedia" timeline of cosmology. For both particle physics and cosmology there is a wealth of web-based information on the history. For any item where you wish to know more we suggest that you start by trying a web search.

Early recognitions about the Universe came from the discipline of observational astronomy. Cosmology grew and developed as astrophysics developed as a science. Applying known laws of physics to interpret individual types of observed astronomical objects soon led to the attempt to also interpret overall patterns in those observations from known physics. The recognition that an expanding Universe was a cooling Universe and hence at early times was populated with highly energetic particles brought a link to particle and nuclear physics knowledge. Modern cosmology tests and refines theories based on a match between a wealth of observational data and the output from computer models of the evolution of the Universe that build in the known physics. Both the observations and the modeling have progressed greatly in recent years.

## Relevant Nineteenth-Century Developments

### 1797–1804

**Atoms:** Joseph-Louis Proust carried out a series of experiments showing that the mass fraction of oxygen combined with a given metal after burning had only certain defined values. He also showed that chemical compounds always contain the same mass ratios of their elemental constituents. In 1803 John Dalton provided an interpretation of Proust's results in terms of atoms.

## 1815

**Atoms:** William Prout hypothesized that the atomic weights of elements are whole-number multiples of the atomic weight of hydrogen. We now see these numbers as a counting of the number of protons plus neutrons.

## 1826

**Stars:** Heinrich Olbers restated the problem (mentioned by others earlier, including by Kepler in 1610) that, if the Universe is infinite, static, and uniformly populated with stars, then the night sky should appear equally bright in all directions. The fact that we see stars in a dark sky is thus commonly known in astronomy as Olbers' paradox. It is an early statement of the fact that observations do not match a static, infinite Universe.

## 1849

**Speed of Light:** Fizeau made a first measurement of the speed of light. His result was $3.15 \times 10^8$ meters per second.

## 1859–1861

**Atomic Spectra:** Kirchhoff and Bunsen measured wavelengths of atomic spectral lines, establishing that spectra are unique to each element. They and others then used spectral analysis to identify new elements.

## 1861–1865

**Electromagnetism:** Maxwell, in a series of papers, described the interrelation of electric and magnetic fields, thereby unifying them into electromagnetism. This work culminated in the now famous Maxwell's equations.

One prediction of these equations is that there are traveling electromagnetic waves in addition to static electric and magnetic fields.

## 1856

**Speed of Light:** Kohlrausch and Weber measured the constant $c$ that appears in Maxwell's equations as the speed of propagation of electromagnetic waves. Within the experimental uncertainty they found a number similar to the speed of light. As early as 1857 Maxwell speculated that this suggested that light is such a wave.

## 1867

**Atom Model:** Kelvin proposed a vortex atom, a geometrical structure that was a stable assembly of interlocking vortex rings. Kelvin conjectured that a classification of the knots would yield a classification of the elements. This idea went nowhere!

## 1869

**Atoms:** Mendeleev classified the known chemical elements in a "periodic table" according to atomic mass and chemical properties, with gaps for unknown elements.

## 1874

**Electrons:** G. Johnstone Stoney suggested that there is some basic unit of electricity, for which he coined the name *electron*. He based this on the phenomenon of electrolysis: he saw that the amount of electricity produced was directly related to the number of chemical bonds broken. Thus he also suggested that electrons are contained within atoms, although at the time atoms were thought to be indivisible objects.

## 1875

**Electrons:** Crookes began studying the discharge from a cathode in glass tubes that he made to contain low pressure gas, now known as a Crookes tube. He discovered *cathode rays*, a stream of something that flowed from the cathode and caused fluorescent discharges in certain gases. He guessed this was a stream of particles and suggested that these particles were a fourth state of matter. It was not for some years that their properties were elucidated; we now understood that they are a stream of electrons.

## 1875

**Atoms:** Maxwell noted that atoms must have a substructure with some internal motion possible.

## 1878

**Atoms:** Lorentz, in his inaugural address at the University of Leiden, summarized the best current view of the structure of matter: Matter is subdivided into molecules, which are composed of atoms. Atoms are characterized by their optical properties (spectra).

## 1886–1891

**Light as Electromagnetic Wave:** In a series of experiments, Heinrich Hertz demonstrated the production of radio waves and established that both radio waves and light are electromagnetic waves of different frequencies, thereby verifying this prediction of Maxwell's theory of electromagnetism.

## 1895

**Electrons:** Perrin observed that cathode rays are negatively charged.

## 1897–1899

**Electrons:** J. J. Thomson, in a series of experiments, showed that cathode rays consist of a single type of negatively charged particle. He found the same charge to mass ratio no matter how cathode rays were produced. He concluded that the particles (then called corpuscles) are universal constituents of all atoms, and that their mass is about 1/1000 of that of a hydrogen atom. Thus he discovered what we now call the electron.

1900–1903

**Atom Model:** J. J. Thomson, about 1900, proposed a model for the atom: a number of electrons moving in some vaguely specified positive-charge background of indefinite shape. As late as 1903, Thomson thought that a hydrogen atom contained about 1000 electrons. Lorentz thought that an electron too was an extended charge distribution.

## 1900–1930: Development of Quantum Ideas, Beginnings of Scientific Cosmology

1900

**Quantized Radiation:** Planck, trying to understand the observed spectrum of energy radiated from a black body (any black object at a temperature warmer than its surroundings) suggested that the radiation is quantized; that is, each frequency is present with certain discrete amounts of energy. Planck presented his ideas to an audience of not quite 25 persons. Nobody understood much, except that the final formula fitted measurements that had previously been unexplained. Only a few physicists paid attention to the quantum ideas; there was much confusion as to what they meant. There was no obvious connection made in Planck's treatment between the quantum radiation and the structure of matter.

1905

**Quantized Radiation:** Einstein, one of the few physicists then taking Planck's quantum ideas seriously, proposed that all light consists of discrete energy packets (now called photons) whose energy is proportional to frequency, as Planck had assumed for the case of black-body radiation. Einstein showed how this could explain features of the photoelectric effect, in which light falling on a surface ejects electrons from the surface. Nobody, including Planck, took this idea seriously at first. (Einstein was awarded the Nobel Prize for 1921 for this work; his other 1905 work, on relativity, and his 1915 work on general relativity were still regarded as too speculative.)

1905

**Special Relativity:** Einstein published his theory of special relativity, which refined and developed earlier work, particularly that of Poincaré and Lorentz. He postulated that all observers, in any inertial reference frame, see the same physics, including the same value for the speed of light (or of any other massless particle). This led to predictions that the physics of one frame as seen from another has some surprising properties; since then these properties have been tested and confirmed.

### 1909

**Nuclei:** Geiger and Marsden, in an experiment studying the scattering of $\alpha$-particles from thin foils, observed that a few particles scatter at large angles, some even backward.

### 1911

**Nuclei:** Rutherford realized that the experiments of Geiger and Marsden (carried out under his supervision) suggested the existence of a tiny massive core inside an atom, now called the nucleus. Rutherford's idea had decidedly mixed acceptance. It was violently opposed by J. J. Thomson.

### 1912

**Red-Shift:** Vesto Silpher showed that all "stellar nebulae" (objects later seen to be distant galaxies) have a pattern of atomic absorption lines that is red-shifted; thus a Doppler shift interpretation implies that they are in motion away from us.

### 1912

**Stars:** Henrietta Leavitt, working at the Harvard College observatory, first as a volunteer, later paid 30 cents per hour, showed that Cepheid variable stars have a fixed relationship between their observed period and their luminosity.

### 1913

**Atomic Model:** Bohr constructed a theory of the electronic structure of atoms based on quantum ideas. His theory treated the nucleus as a heavy and very small (effectively point-like) positively charged core and introduced the idea of quantum states (or quantum levels) for any electrons in the Coulomb field of this core, and quantum jumps (or quantum transitions) of electrons between these states as the mechanism that produces the characteristic spectral lines of radiation from atoms. His theory explained for the first time the known energy patterns of these lines.


## 1915

**Symmetry:** Emmy Noether published a mathematical work that showed a deep relationship between symmetries, or invariances, in the physical equations and conservation laws.


## 1915

**General Relativity (GR):** Einstein published his general theory of relativity, showing that mass causes curvature in space-time. His theory explains a detail of the orbit of Mercury, known as the precession of the perihelion, which had been a puzzle up to that time.


## 1917

**Cosmology from GR:** Einstein applied his theory of general relativity to the Universe as a whole. He found that it gave a time-varying Universe but that he could add to it an additional parameter, now known as the cosmological constant, which can be chosen to balance the overall gravitational effect of mass so that the Universe is static, even when it contains an overall matter density. (Later he called this his biggest mistake.)


## 1918

**Distance to a Star:** Shapley showed how Leavitt's (1912) relationship allows one to determine the distance to a Cepheid variable star.


## 1919

**Nuclei:** Rutherford observed the first nuclear transmutation, $^{14}_{7}N + \alpha \rightarrow {}^{17}_{8}O + p$. This process provided the first evidence that the nucleus of the hydrogen atom is also a constituent of other nuclei. Two years later, Rutherford postulated that it is a fundamental particle and named it the proton. Rutherford also made a first estimate of nuclear size.

### 1919

**General Relativity Test:** Eddington observed the light from a star passing behind the sun during a solar eclipse and demonstrated the gravitational bending of its path predicted by Einstein's general relativity.

### 1921

**Strong Nuclear Force:** From studies on alpha-hydrogen scattering, Chadwick and Bieler concluded that some kind of strong force (not following the $1/r^2$ force law of electric forces) exists inside the nucleus.

### 1922

**Cosmology from GR:** Alexander Friedmann showed that Einstein's theory of general relativity includes possible solutions corresponding to an ever-expanding universe, or one that expands but then at a later time contracts. The history depends on both the mass density and the cosmological constant. Histories with an expanding universe with zero cosmological constant are possible.

### 1923

**Quantum Radiation:** Compton discovered the particle nature of X-rays. Like photons, they are quanta with definite energy-frequency and energy-momentum relationships.

### 1924

**Wave-Particle Duality:** de Broglie introduced wave-particle duality, conjecturing that all particles have wave-like properties and all radiation has particlelike quantum properties.

1924

**Distance to a Galaxy:** Hubble applied the technique of Leavitt (1912) and Shapley (1918) to calculate the distance to a faint Cepheid variable star observed with the Mount Wilson telescope in the Andromeda nebula, showing that it was well outside our galaxy, thus showing that nebulae were in fact distant galaxies. Until that time the astronomical community had been divided on this point, many thinking that these nebulae were bright gaseous regions within our galaxy.

1925

**Exclusion Principle:** January—Pauli formulated the exclusion principle: no two electrons can occupy the same state in an atom. He recognized that his rule required an unexpected parameter (that can take only two values) labeling the quantum states of electrons in atoms. (This was later recognized as the two possible spin orientations.) The 1945 Nobel Prize recognized this work. With this rule and Bohr's sets of quantum states the light element patterns of the periodic table of the elements can be understood.

1925

**Conservation Laws in Atomic Processes:** April—Bothe and Geiger demonstrated that energy and momentum are conserved in individual atomic processes. Since the introduction of quantum ideas had thrown out many standard ideas of classical physics it was up to this point not completely clear whether these important physical laws applied in the quantum world or not.

1925

**Electron Spin:** October—Goudsmit and Uhlenbeck introduced spin as a new electron attribute, assigning intrinsic angular momentum of $1/2\ \hbar$ to the electron, thereby explaining the two-valued parameter that Pauli had noted was needed to categorize electron states in an atom.

1925–1926

**Quantum Formalism:** Heisenberg invented matrix mechanics and within months Schrödinger developed wave mechanics. Both are mathematical formulations of quantum theory. They allow one to calculate the possible states of a quantum system, such as the states of electrons in atoms. The results correct and improve Bohr's simple model. Soon after, Dirac showed that the two theories are equivalent within a more general framework. Born gave an interpretation of the wave function that appears for each state in the Schrödinger formulation, showing that it determines the probability distribution for the location of the electron in that state. (Also at this time G. N. Lewis introduced the name photon for a light quantum.)

1927

**Proton Spin:** Dennison, by analyzing hydrogen spectral lines, determined that the spin of the proton is $1/2\ \hbar$, the same as that of the electron.

1927

**Beta-Decay Spectrum:** The spectrum of energies of electrons, emitted by certain nuclei in the process known as beta decay, was shown to have a continuous range of energies. At the time, it was unclear whether these electrons were of nuclear or atomic origin. Since both the atom and the nucleus were, by this time, recognized to have discrete energy levels, it was hard to understand these results without giving up the conservation laws of energy, momentum, and angular momentum, which had been shown to work in other atomic processes (see "1930" for the answer).

1927

**Quantum Uncertainty Principle:** Heisenberg formulated the uncertainty relations, showing that in quantum theory it is impossible to make simultaneous, arbitrarily precise, measurements of both a particle's momentum in some direction and its coordinate in that direction. Precise measurement of one causes a spread of possible values for the other.

1927

**Parity:** Wigner introduced the concept of parity of quantum states (a consequence of left–right symmetry). Parity is an additional label that defines states by the properties (even or odd) of their wave functions under reflection of all three spatial directions about the origin. He showed that this property is a consequence of an invariance of the equations describing the interactions under such a change of coordinate definitions. This extra label explained a pattern that had been noted in atomic spectra (Laporte's rule), and thus was rapidly accepted.

1927

**Cosmology from GR:** Le Maitre stressed that Einstein's general relativity theory allows for an expanding universe and suggested that this is a correct cosmology, but did not offer any evidence to support this view.

1928

**Quantum Formalism:** Dirac derived a new equation that combines quantum mechanics and special relativity to describe the electron. His reasons for rejecting the Schrödinger equation were not correct, but his equation correctly predicted the electron's magnetic properties. (Later it was recognized that Dirac's equation applies for all half-integer spin particles while Schrödinger's applies for integer spin particles.) In the following years it was recognized that Dirac's equation also requires the existence of corresponding oppositely charged particles—now called antiparticles. Dirac and Schrödinger were awarded the Nobel Prize in 1933.

1929

**Particles:** Dirac proposed (incorrectly) that the positively charged particles required by his equation are protons. Meanwhile Born, after learning about the Dirac equation said, "Physics as we know it will be over in six months."

1929

**Red-Shift–Distance Relationship for Galaxies:** Hubble used his data on distances to galaxies to obtain a linear relationship between red-shift and distance, and thus evidence that the Universe is expanding. His relationship was a very crude fit to his data, which were of mixed quality. This inspired guess was later verified over a much larger range of distances.

1930

**Summary:** A general consensus of physicists at this time would include quantum mechanics and special relativity. However, it would recognize just three fundamental particles: protons, electrons, and photons. General relativity and its implications for cosmology were still widely regarded as abstract and speculative.

## 1930–1950: New Particles, New Ideas

1930

**Neutrino:** Pauli in letters and private conversations suggested that an additional new type of particle, the neutrino, must be being produced to explain the continuous electron spectrum for beta-decay. If three particles are produced by the transition, only the sum of their energies is discrete.

1931

**Antimatter (Positron Prediction):** Dirac finally accepted that the positively charged particles required by his equation are new objects. They must be exactly like electrons, in particular they have the identical mass, but positively charged. This is the first example of antiparticles.

1931–1932

**Antimatter (Positron Discovery):** Anderson observed positively charged particles of about the same mass as the electrons produced by cosmic rays. He called them positrons and suggested a renaming of electrons as negatrons. Pauli and Bohr at first did not believe that these were Dirac's positive particles but further data, including electron-positron pair production, soon provided good evidence of that.

1932

**Neutron:** Discovery of the neutron, with its spin determined to be 1/2 $\hbar$. Chadwick was awarded the Nobel Prize in 1935 for his work demonstrating the existence of the neutron. If a neutron is as elementary as a proton, nuclei can then be understood as composed of protons and neutrons. The questions of the mechanisms of nuclear binding and decay become primary problems for physics. Essentially all of particle physics was discovered in the ensuing attempt to understand nuclear interactions.

1931

**Data Fit Expanding Universe Theory:** George Le Maitre pointed out that Hubble's red-shift to distance relationship supports the expanding Universe theory. He was the first to note that such a theory implied a beginning to the history of the Universe, the moment now called the Big Bang. He noted that such a cosmology has a starting time, when the Universe begins its expansion from an infinitely dense state. Le Maitre, an ordained Jesuit priest as well as a Ph.D. physicist, considered this starting moment to be the moment of creation.

1933–1934

**Beta Decay and Neutrinos:** Fermi proposed a theory of beta decay based on Pauli's neutrino hypothesis. This was the first introduction of an explicit additional type of interaction, now called weak interactions, also the first explicit use of neutrinos, and the first theory of processes in which a fundamental particle changes type ($n \rightarrow p + e^- + \nu_e$). The introduction of a new interaction, in addition to the familiar electromagnetic and gravitational interactions, was a very radical step. The idea that the electron (and the neutrino) produced in nuclear beta decay are in no way present in the nucleus before the decay was also radically new. Fermi's approach was pragmatic; he found a formulation that gave a fit to data if one assumed there was an unseen neutrino being produced, and he tested it with further experiments.

### 1933–1934

**Strong Nuclear Force, Meson Theory:** Yukawa introduced a new idea about the nature of strong nuclear forces. Combining relativity and quantum theory, Yukawa described nuclear interactions as mediated by an exchange of a new type of particle between protons and neutrons. (He tried at first to use electrons but found that did not fit the observations.) From the size of the nucleus, which gives the range of the new interaction, Yukawa concluded that the mass of these conjectured particles (mesons) is about two hundred electron masses. This is the beginning of the meson theory of nuclear forces. The Nobel Prize was awarded to Yukawa in 1949.

### 1933

**Homogeneous and Isotropic Universe:** Edward Arthur Milne introduced a cosmological expanding universe model without general relativity which has not survived later tests. However, in this work, this mathematical physicist was the first to clearly articulate the symmetry principle that today goes by the name he gave it: the cosmological principle. This is the concept that no object or region occupies any special place in the Universe, which leads to the conclusion that the Universe, at least on the large scale, looks much the same no matter from what point it is viewed. The Universe, averaged over a large enough scale, is both homogeneous (the same everywhere, no preferred place) and isotropic (the same looking out in any direction, no preferred direction). This principle is a basic part of all modern theories of cosmology.

### 1933

**Dark Matter:** Astronomer and astrophysicist Fritz Zwicky pointed out that certain types of galaxies must contain ten to one hundred times more mass in some form other than that which makes visible stars, to explain gravitational pull on the outer stars around the centers of their galaxies.

This is the first suggestion of the presence of what is now known as dark matter.

1937

**Unexpected Particle—Muon:** A particle of mass about two hundred times the electron mass was discovered in cosmic rays by Neddermeyer and Anderson and, separately, by Street and Stevenson. First thought to be Yukawa's meson, it was later recognized that it must be another entirely new and unexpected type of particle, a heavy electron-like particle, now called a muon (see 1946–1947).

1938

**Conservation Laws to Prevent Proton Decay—Baryon Number:** Stückelberg observed that protons do not decay although many lighter particles exist. The stability of the proton cannot be explained in terms of energy, momentum, or electric charge conservation. He proposed an independent conservation law; in contemporary language this is the conservation law of baryon number or quark number. In 1949 Wigner gave a more explicit formulation, and observed that the decay $p \rightarrow e^+ + \gamma$ could satisfy all other conservation laws but does not occur.

1941

**Muon Decay:** The first measurement of the muon lifetime (still thought to be Yukawa's meson). (It decays to an electron, a neutrino, and an antineutrino.)

1941

**Nucleons:** Möller and Pais introduced the term *nucleon* as a generic term for protons and neutrons.

1946–1947

**Muon:** It was realized that the cosmic ray particle thought to be the Yukawa meson cannot be any such thing because its tracks in emulsions show that it does not interact strongly as it passes through matter. It is instead a particle with no strong interactions, just like an electron, except that it is more massive and thus unstable. The muon ($\mu$), the first particle of the second generation to be found, was completely unexpected. I. I. Rabi commented: "Who ordered that?" The term *lepton* was introduced as a generic name for objects that do not interact strongly, namely, the electron, the muon, and the neutrinos. Correspondingly, the generic term *hadron* was introduced for particles that do have strong interactions, such as the proton and the neutron, and Yukawa's (still undiscovered) meson.

1947

**Mesons Discovered:** Powell and collaborators, using sensitive nuclear emulsions exposed to cosmic radiation, discovered another type of particle with mass a little greater than a muon, which does interact strongly. It decays into a muon and a neutrino. This was identified as Yukawa's strongly interacting meson, now called the pion.

1947

**Quantum Formalism—QED:** Calculation procedures were developed for quantum electrodynamics (QED), the relativistic quantum field theory generalization of Dirac's equation. Feynman played a major role in these developments; the diagrams he introduced as a device to keep track of his calculations influenced the way all later physicists think about and calculate quantum processes.

1948–1960s: The Advent of Accelerator Experiments— The Particle Explosion; Implications of Expanding Universe Explored

1948

**Accelerator:** The first artificially produced pions were observed at the Berkeley synchrocyclotron.

1948

**Nuclear Synthesis:** Alpher and Gamow recognized that nuclei for all elements can be synthesized starting from protons and neutrons by a sequence of fusion and neutron capture reactions. They found some problems in matching the observed ratios of different elements to a model of production via cosmological processes. (They invited Hans Bethe to be a coauthor on their paper, making its author list into a play on words with their three names $(\alpha,\beta,\gamma)$.)

1948

**New Steady-State Universe Proposed:** Noted astronomers Hermann Bondi, Thomas Gold, and Fred Hoyle suggested a steady-state theory of cosmology capable of explaining Hubble's data, as a counter to the prevailing view of an expanding Universe. Hoyle in a radio interview called the expanding Universe theory the Big Bang, intending to mock it, but the name became accepted as a good description of the idea. The steady-state model has been ruled out by later data, which clearly support an expanding Universe.

1948

**Cosmic Microwave Background Radiation:** George Gamow suggested that the photons from a hot early Universe should be present today as a black-body spectrum of microwave radiation. This idea was developed and refined by Alpher and Hermann in a 1950 paper.

1949

**Mesons as Composite Particles:** Fermi and Yang suggested that a pi-meson $(\pi)$ is a composite structure of a nucleon and an antinucleon. The idea that so light a particle could be composite was quite radical in 1949. (Nowadays it is seen as a composite of a quark and an antiquark.)

1949

**Unexpected New Mesons:** A new type of meson, now called $K^+$, was discovered by the Bristol group (Powell and collaborators) via the emulsion tracks showing its decay, $K^+ \rightarrow \pi^+\pi^-\pi^+$. (As early as 1944, Leprince-Ringuet and L'héritier had seen less substantial evidence for the $K^+$.)

## 1950

**Mesons with Zero Electric Charge:** Panofsky, Steinberger, and Steller showed that nucleon-nucleon scattering experiments require that there are pions with zero electric charge as well as those with positive and negative charges. They found evidence for the production of neutral pions via their decay to two photons.

## 1951

**More Particles—Mesons and Baryons:** Rochester and Butler discovered two new types of particles in tracks produced in a bubble chamber in processes initiated by cosmic rays. They looked for V-like tracks, which can be interpreted as the diverging tracks of two charged particles produced by the decay of a parent electrically neutral particle (which leaves no track). The mass of the parent particle can be deduced by reconstructing its energy and momentum from those of the (known-mass) products. The new electrically neutral particles were at first called V particles. The two types found in this way are now called $\Lambda^0$ (which was seen by its decays, $\Lambda^0 \rightarrow$ proton $+ \pi^-$) and $K^0$ (which was seen to decay via $K^0 \rightarrow \pi^+ + \pi^-$). The $\Lambda^0$ was the first particle more massive than a neutron to be discovered.

## 1952

**New Detector Technology:** Glaser invented the bubble chamber, the first detector that allowed large-volume studies of particle tracks. Photographic images of the bubbles along particle tracks in this chamber were scanned by teams of scanners (mostly women in a time when physics was essentially an all-male field) to find particle production or decay sites.

## 1952–1953

**More Baryons:** In the mid-1950s, the first evidence was seen for what we now call the Delta particles. These were seen as resonances produced in photon-proton scattering, and decaying to pion plus neutron or proton. The initiating high-energy photons were produced by 300 MeV electron-synchrotron accelerators at Berkeley, Cornell, and MIT and soon a slightly higher-energy machine at Caltech. By 1953, with pion-proton scattering data from the Chicago cyclotron experiments as well, such particles had been seen with four different charge states, $\Delta^{++}$, $\Delta^+$, $\Delta^0$, $\Delta^-$. Over the following years the data on these and many other resonances have been steadily improved. Their decays are so rapid that they are not directly observable; instead, they appear as resonant peaks in the rate of events as a function of energy.

## 1953–1960s

**More Particles:** A veritable particle explosion took place, with discovery of many new elementary particles, both meson types and baryon (proton-like) types. Classification of particles and their decays became a major activity. Particle decays were classified on the basis of their half-life, and the types of particles produced in the decay, into three groupings. Particles that decay by strong interaction processes, into less massive mesons and baryons, such as the Delta decays described above, have half-lives of order $10^{-24}$ seconds; electromagnetic decays, which produce one or more photons, have half-lives between $10^{-19}$ and $10^{-16}$ seconds; decays via weak processes typically produce neutrinos and have half-lives of $10^{-13}$ seconds and longer. The slower decays can generally be observed only when no more rapid decay path is possible. Particles could also be classified by spin (fermions with half-integer spin, or bosons with integer spin) and by whether or not they participate in strong interactions (hadrons or leptons). By 1964 over one hundred elementary particle types had been identified.

## 1953

**Lepton Conservation Laws:** The law of conservation of lepton numbers was first stated in a paper by Konopinski and Mahmoud, noting that hadron weak decays that produce a charged lepton must also produce a matched neutrino (or antineutrino). This law extends to lepton weak decays, in which case two neutrinos are produced, one matched to the decaying lepton and one (anti)matched to the produced lepton.

1953

**Strangeness Conservation in Strong and Electromagnetic Decays:** Gell-Mann and Nishijima introduced a new particle attribute, called strangeness, to explain the disparity between the copious production of $\Lambda$'s and $K$'s and their relatively slow decays. The idea is that particles with strangeness decay only by relatively slow components of the weak interactions, but can be produced by strong or electromagnetic processes, always in pairs of particles with opposite sign of strangeness. The modern interpretation of this property, also developed by Gell-Mann, is that the $\Lambda$ contains a strange quark, and the $K^+$ contains a strange antiquark ($\Lambda = uds$ and $K^+ = u\bar{s}$). A strange quark and its antiquark can be produced together in a strong or electromagnetic interaction, but each decays only by a rare type of weak interaction.

1953–1957

**Nucleon Substructure:** Experiments that bombard nuclei with high-energy electrons showed that the electric charge inside protons has a distribution of varying density. Even neutrons were seen to have some internal charge density distribution. This internal structure of protons and neutrons is something of a puzzle as they were thought to be elementary particles, which implicitly meant structureless.

1954

**Quantum Formalism Allowing Charged Spin-1 Particles:** Yang and Mills developed the mathematical formulation of a new class of theories for spin-1 particles, called non-Abelian gauge theories. This type of theory now forms the basis of the Standard Model. Initial attempts to use the Yang-Mills formulation tried to apply it to the then known spin-1 mesons, such as the rho-mesons, which are massive, but failed, in large part because the mathematics describes only massless particles.

1954–1957

**Antimatter:** The Berkeley Bevatron, a 6.2 GeV (then called BeV) machine, began operation. This machine was designed to have high enough energy to create proton–antiproton or neutron–antineutron pairs. Segre and Chamberlain led the experiment that discovered the antiproton there in 1955 (Nobel Prize 1959). In 1956, in this same facility, experiments by Cork, Lambertson, Piccioni, and Wenzel obtained evidence for the antineutron.

## 1956

**Neutrinos Detected:** After years of effort, experiments by Clyde Cowan and Fred Reines with a detector close to the Savannah River Nuclear Reactor showed the first direct evidence for the existence of neutrinos by recording events initiated in their detector by neutrinos produced in the reactor. In 1995 the Nobel Prize finally recognized the pioneering nature of this work. Reines shared the prize that year with Martin Perl; Cowan was no longer living.

## 1957

**Neutrino Mass:** Bruno Pontecorvo suggested that neutrinos are not necessarily massless, and that, if they have masses, it is possible that the neutrino type (in this case neutrino versus antineutrino) observed at capture could be different from the type produced, due to a phenomenon now called neutrino oscillation.

## 1956–1957

**Parity Is Not a Symmetry of Weak Decays:** Lee and Yang observed that the parity (P) invariance of basic laws (the property that they do not change form when all directions are reflected) and charge-conjugation (C) invariance (the particle–antiparticle symmetry) had never been checked in weak interactions. They suggested that weak interactions may not possess these symmetries, though it is known that strong and electromagnetic interaction laws do. Experiments on beta decay by Wu *et al.* and muon decay indeed show violations of P and C invariance. The Nobel Prize was awarded to Lee and Yang in 1957. They also showed how Fermi's weak interaction theory can be rewritten to match this observation.

1957

**Weak and Electromagnetic Theories Related:** Schwinger proposed a theory that attempted to unify the weak and electromagnetic interactions. His theory was incomplete, but contained some of the basic ideas that are now part of the Standard Model.

1957

**Most Nuclei Formed in Stars:** Burbridge, Burbridge, Fowler, and Hoyle showed that all elements (except the very lightest, hydrogen to lithium) are formed via nuclear fusion processes in the core of stars. (Supernova explosions are required to account for elements more massive than iron.)

1957–1959

**Weak Interaction Mediated by Massive Spin-1 Particles:** Separate papers by Schwinger, Bludman, and Glashow suggested that all weak interactions are mediated by charged, very massive spin-1 bosons, later called $W^+$ and $W^-$. The name $W^{\pm}$ was first used by Lee and Yang in 1960.

1961

**Patterns of Particles:** Gell-Mann exploited the patterns of particles of similar mass and spin, but differing charge and strangeness, to create a classification scheme (based on the group SU(3)) now called flavor symmetry, that encompasses all then known particles. The scheme predicts a new particle type, $\Omega^-$, a spin-1/2 hadron with strangeness −3 and charge −1. The discovery of this particle soon thereafter gave great support to this idea. For his earlier work on strangeness, this classification scheme, and the later work on quarks (see 1964), which followed directly from this idea, Gell-Mann was awarded the Nobel Prize in 1969.

1962

**Two Neutrino Types:** Experimental verification that there are two distinct types of neutrinos ($\nu_e$ and $\nu_\mu$) was obtained. The experiment showed that the neutrinos produced in association with a muon (in pi-meson decays) do not cause the interactions that would be expected if they were the same as the neutrinos produced in nuclear beta decays. The experimenters, Leon Lederman, Melvin Schwartz, and Jack Steinberger, were awarded the Nobel Prize in 1988 for this work. Most physicists take this to indicate that neutrinos are massless, because any mass would be likely to confuse these separate identities via neutrino oscillation. In fact, all the experiment can really tell us is that such an oscillation effect is extremely rare over the distance between the neutrino source and the detector, implying that neutrino masses—if not zero—are very small.

## 1964–1973: Formulation of the Modern View of Particles and the Universe

### 1964

**Particle Substructure—Three Types of Quarks:** Quarks were first tentatively introduced by Gell-Mann and independently by Zweig. Quarks give a basis in terms of the particle structure for the classification scheme proposed earlier by Gell-Mann (see 1961). All mesons and baryons are composites of three species of quarks and antiquarks, now called $u$, $d$, and $s$, of spin-1/2 $\hbar$ and with electric charges (2/3, −1/3, −1/3) in units where the proton charge is +1. The similarly charged $d$ and $s$ quarks are distinguished by the fact that the $s$ carries the strangeness quantum number −1, while the $d$ has zero for this quantity. These fractions of a proton or electron charge had never been observed. The introduction of quarks was generally treated more as a mathematical explanation of flavor patterns of particle masses (see 1961) than as a postulate of actual physical objects. However, the great simplification obtained, from over one hundred "fundamental" hadron-type particles to just three, made the idea very attractive. Later theoretical and experimental developments allow us to now regard the quarks as real physical objects, even though they cannot be isolated.

### 1964

**CP (Matter–Antimatter) Symmetry Is Not Exact:** Jim Christensen, Jim Cronin (Nobel Prize 1980), Val Fitch (Nobel Prize 1980), and Rene

Turlay found, much to everyone's surprise, that the long-lived neutral K-meson sometimes decays to two pions. This decay would be forbidden by an exact matter–antimatter symmetry in the laws of physics. All particle theories known at the time had such a symmetry. Cronin's Nobel lecture tells of the skepticism with which their results were met.

## 1964

**Fourth Quark Type Suggested:** Stimulated by the repeated pattern of leptons, several papers suggested a fourth quark carrying another flavor to give a similar repeated pattern for the quarks, now seen as flavor generation patterns. Very few physicists took this suggestion seriously at the time. Glashow and Bjorken coined the term charm for the fourth ($c$) quark.

## 1964

**Primordial Nucleosynthesis Depends on Number of Neutrino Types:** Fred Hoyle and Roger Taylor, developing simulations of cosmological production of light element nuclei, showed that their results for the ratio of helium to hydrogen depended on the number of light neutrino species, since this affects the rate of expansion and cooling and hence the ratio of neutrons to protons at the time that the Universe is cold enough for the nuclei that form to be stable in typical collisions. This is called the time of nucleosynthesis.

## 1965

**Quark Color Charges:** Color charge, an additional degree of freedom, was introduced as an essential property of quarks by Greenberg, and by Han and Nambu. Each quark flavor type can carry each of three colors. Antiquarks carry anticolors. (Color is a name for the strong charge attribute; it is not related to visible colors.) All observed hadrons must be color neutral. This is achieved for the three-quark structure of baryons, and likewise the three-antiquark structure of antibaryons, as well as the quark–antiquark structure of mesons. Color charge was first introduced as a mathematical device to explain the patterns of light baryon types;

later it was understood to be the charge associated with strong interactions that bind the quarks to form the color-neutral hadrons.

## 1965

**Cosmic Microwave Background Radiation:** Arno Penzias and Robert Wilson observed signals in a radio telescope that at first they thought was noise in the system, but which they could not eliminate. Conversations with their colleagues, Dicke, Peebles, Roll, and Wilkinson, led to the recognition that they had observed the cosmic black-body radiation, light from the early Universe now red-shifted to microwave wavelengths. Penzias and Wilson were awarded the 1978 Nobel Prize in physics for this discovery.

## 1966

**Quarks Not Accepted:** The quark model was accepted rather slowly because quarks were not directly observed. In Gasiorowicz's popular textbook on particle physics published in 1966, quarks are not mentioned.

## 1966

**Primordial Nucleosynthesis:** Jim Peebles showed that, to within the accuracy of his calculations for nucleosynthesis, the expanding Universe cosmology gives the correct ratio of primordial helium (not made in stars) to hydrogen.

## 1967

**Weak and Electromagnetic Theories Related:** Weinberg and Salam separately proposed versions of a theory that unifies electromagnetic and weak interactions. It was of the Yang-Mills type (see 1954). The theory requires the existence of a neutral weakly interacting boson (now called the $Z^0$) as well as the $W^+$ and $W^-$. This particle mediates a weak interaction that had not been observed at that time. They also predicted an additional massive boson called the Higgs boson, needed to modify the Yang-Mills approach to include particle masses. This particle has yet to be observed. Their idea was mostly ignored; from 1967 to 1971, Weinberg's and

Salam's papers, now regarded as the first suggestion of an essential part of the Standard Model, were rarely quoted. One reason for this was that no one knew how to do calculations for this theory.

## 1967

**Primordial Nucleosynthesis:** Robert Waggoner, William Fowler, and Fred Hoyle extended the analysis of Peebles (see 1966) to show that the hot Big Bang model gives the correct abundances of deuterium and lithium, in addition to helium, relative to hydrogen, within the uncertainties of the simulation.

## 1967

**Cosmology—Matter–Antimatter Imbalance:** Andrei Sakharov formulated the three properties a theory must have to allow the development of a matter–antimatter imbalance, starting from a hot Universe with equal populations of particles and antiparticles. For this to occur, the laws of nature must include baryon- or lepton-number-changing processes, C and CP symmetries must be broken, and the process by which the imbalance is created cannot be governed by thermal equilibrium conditions.

## 1968

**Neutrinos from Sun:** Ray Davis (Nobel Prize 2002) contrived and built an underground experiment to detect the neutrinos produced by the nuclear processes in the Sun. This experiment was first proposed by Ray Davis and John Bahcall in 1964 and first results were reported in 1968. Bahcall was the theorist who calculated both the expected neutrino flux from the Sun and the expected capture rate for neutrinos in the chlorine detector. The number of neutrinos observed was about one-third that expected from models of how the Sun shines, and for over twenty years Davis continued the work with similar results. Most physicists thought that either the experiment was inaccurate or the solar model was inaccurate. Later measurements have shown that both experiment and solar model were remarkably accurate. What was not right was the particle theory of massless neutrinos. It took forty years for this to be established, which explains the long gap between this experiment and Davis's Nobel Prize.

1968–1969

**Quark Substructure of Nucleons Observed:** Observations at SLAC (Stanford Linear Accelerator Center) indicated that in inelastic electron-proton scattering the electrons appear to be bouncing off small dense objects inside the proton. Bjorken and Feynman analyzed these data in terms of a model of constituent particles inside the proton, without using the name quark for the constituents. Taylor, Friedman, and Kendall were awarded Nobel Prize in 1990 for this experimental evidence for quarks.

1969

**Cosmology—Horizon Problem:** Charles Misner formalized the statement of a problem for the standard Big Bang theory, first recognized by Alpher, Herman, and Follin in 1953, known as the horizon problem. Since any form of information travels at the speed of light or less, in a finite life of an expanding Universe only a limited region could have been in thermal contact at any time in the past. The current homogeneity of the Universe requires that large regions were in the same thermal state at the same early time, regions that, if the Universe expands only as a power law with time, would never have been in thermal contact. This is known as the horizon problem.

1969

**Cosmology—Flatness Problem:** Robert Dicke cleanly formulated another problem, known as the flatness problem. Observations show that space-time is close to flat. In an expanding Universe this requires that the average energy density be close to a critical value. Since mass density decreases as the Universe expands (unless it is exactly at the critical value), if the energy density is dominated by matter (mass) density, to be close to the critical value today the Universe must have been extremely close to that value in the past. A good theory should explain why this must be so, or even give a reason why the value is exactly critical. Astronomers were aware that the observed matter gives only a small fraction of the critical energy density; most cosmologists expected that the difference would be made up by dark matter.

## 1970

**Fourth Quark Needed to Explain Weak Interaction Data:** Glashow, Iliopoulos, and Maiani pointed out that a fourth type of quark is needed in the context of the Weinberg-Salam type theory of weak and electromagnetic interactions (see 1967). A fourth quark allows a theory that has flavor-conserving $Z^0$-mediated weak interactions but not flavor-changing ones.

## 1971–1972

**Quantum Formalism:** Gerard 't Hooft (then a graduate student) and Martinus Veltman (his advisor) developed calculational tools that allow the Weinberg-Salam theory to be treated beyond a first-order approximation. Recognition that this theory is well behaved (compared to Fermi theory, which gives nonsense beyond the first-order approximation) led to growing interest in it. 't Hooft and Veltman were awarded the 1999 Nobel Prize for this work.

## 1972

**Strong Interaction Theory:** A definite formulation of a quantum field theory of strong interactions was proposed. This theory of quarks and gluons (now part of the Standard Model) is similar in mathematical structure to quantum electrodynamics (QED); hence the name quantum chromodynamics (QCD). It is also a Yang-Mills-type gauge theory, but with multiple charge types, known as color charges. Quarks are real particles, carrying a color charge. Gluons are massless quanta of the strong interaction field and, in addition, they also carry color charges. This strong interaction theory was first suggested by Fritzsch and Gell-Mann.

## 1973

**Weak Interaction, Z-Boson-Mediated Processes Observed:** Spurred by a prediction of the Glashow-Illiopoulos-Maiani theory (see 1970), members of the Gargamelle collaboration reanalyzed some old data from CERN and found indications of flavor-conserving processes (due to $Z^0$ exchange). Earlier analysis of this data had not really looked for them. Why not? It was known that there were no processes that changed quark type without changing charge, and so physicists assumed that any such effect, even if it conserved quark flavor, would be tiny. Thus it was thought that the search would be too difficult because of background processes (similar-looking events caused by other processes). After the theoretical prediction, the experimenters figured out a way to exclude background events at the level needed to see this effect.

1973

**Strong Interaction Properties:** Politzer, a graduate student at Harvard, and Gross, a young faculty member at Princeton, together with his graduate student Wilczek, discovered that theories such as the color theory of the strong interactions have a special property, now called *asymptotic freedom.* The interactions between quarks become asymptotically weaker at very short distances (or, equivalently, at very high energies), but become very strong at a scale that defines the size of hadrons. The property correctly described the 1968–1969 data (and all later experiments) on the substructure of the proton. This work was awarded the Nobel Prize in 2004.

1973

**CP Violation:** Kobayashi and Maskawa pointed out that a two-generation Standard Model theory automatically gives identical laws of physics for matter and antimatter and thus cannot accommodate the known CP violation, but that a three-generation theory could do so. Most physicists had not yet accepted the two-generation theory, so this work was largely ignored for some years.

1974

**Summary—The Standard Model of Particles and Interactions Emerges:** In a summary talk for a summer conference Iliopoulos summarized, for the first time in one single report, the view of physics now called the Standard Model. Only particles containing $u$, $d$, and $s$ quarks were known, with the charm quark predicted by the theory, but particles containing it not yet discovered. Most physicists were skeptical about the need for charm, but Glashow declared at an international conference in April that if charm is not found within two years, he would "eat my hat."

## Two Standard Models Emerge—Particles and Cosmology

### 1974

**Fourth Quark Type Discovered—Charm:** November—An extremely narrow, high peak in event rate, signaling production of a new type of particle, was discovered by Richter and his collaborators in the reaction $e^+ + e^- \rightarrow$ hadrons at SLAC (they called the particle $\psi$). The same object was discovered by Ting and his collaborators at about the same time at Brookhaven in a proton experiment and called $J$. In fact, they had found it earlier but kept it secret while they made various checks to understand what they had seen. The two discoveries were announced on the same day, and so are given equal credit. There was a great deal of excitement about the discovery of this new particle of mass 3.1 GeV, now called the $J/\psi$. There were many speculations about what the $J/\psi$ is. The most favored idea was that it is a $c\bar{c}$ state. This interpretation made several predictions, among them that states of $c\bar{u}$ and $c\bar{d}$ (D-mesons) must also exist and be produced at slightly higher energies. Eventually all these predictions were confirmed (see 1976). The 1976 Nobel Prize was shared by Richter and Ting for this discovery.

### 1976

**Mesons with Charm:** A neutral charmed meson, called D-zero ($D^0$), was found by Goldhaber, Pierre, and collaborators in data from the same experiments at SLAC as had seen the $J/\psi$ particles. The observed decay was $D^0 \rightarrow K^- + \pi^+$, as predicted for a charmed meson made from $c\bar{u}$. It was produced in combination with another such particle, anti-D-zero ($\overline{D}^0$), made from $u\bar{c}$, which decays to $K^+ + \pi^-$. Thus these mesons contain charm or anticharm but not both. This dramatically matched the theoretical predictions and confirmed the interpretation of the $J/\psi$ as a $\bar{c}c$ state. (Hence Glashow did not have to eat his hat.)

### 1976

**Another Unexpected Lepton—Tau:** A new charged lepton called $\tau$ (tau), with a mass of about 1.78 GeV, was detected by Perl and collaborators at SLAC. The production of particle–antiparticle pairs of this lepton at almost the same energy as the charm quark and antiquark production threshold was the cause of some of the initial confusion in the interpretation of the $J/\psi$ discovery. Only after the discoveries of the $\tau$-lepton and the D-mesons were sorted out were the data completely understood. This was a strange repetition of history: just as the discovery of a first-generation meson (pion) was confused by the unexpected appearance of a second-generation lepton (muon), so the interpretation of the second-generation mesons ($J/\psi$ and $D$) was confused by the appearance of an equally unexpected third-generation lepton ($\tau$).

### 1976

**Third Neutrino Type Suggested:** The fact that the $e$- and $\mu$-leptons possessed different associated neutrinos strongly suggested that the $\tau$-lepton also has a distinct associated light neutrino, giving six leptons in all. This was a generally accepted idea. The fact that tau-type neutrinos are distinct from muon and electron types has since been confirmed.

### 1977

**Fifth Quark Type:** particles, called Y (upsilon), containing yet another quark (and its antiquark), were discovered by Lederman and collaborators at Fermilab. It was called the bottom quark and carries charge $-1/3$. It gave added impetus to the search for a sixth quark (top), so that the number of quarks would equal the number of leptons and the third repeat of the pattern of particles in the Standard Model (third generation) would be complete.

1978

**Weak Interaction—Confirmation of Z-Boson Effects:** The effect of $Z^0$-mediated weak interactions was clearly observed in the scattering of polarized electrons from deuterium, in an experiment at SLAC led by Prescott and Taylor. The experiment qualitatively and quantitatively confirmed a key prediction of the Standard Model.

1979

**Nucleosynthesis Counts Neutrinos:** Refinements in nucleosynthesis modeling, plus improved data, narrow the window for theories and data to match. Astrophysicists argued that their data were compatible with around two to four types of light neutrinos active in the early Universe, but not more than four.

1979

**Strong Interaction—Evidence for Gluons:** PETRA, a colliding-beam facility at Hamburg, studied events with clusters of hadrons, seeking evidence for gluon-induced effects predicted in QCD. The reaction $e^+ + e^- \rightarrow q + \bar{q}$, with subsequent production of additional quark–antiquark pairs, yields two clusters of hadrons moving in the directions of the initial quark and initial antiquark. A smaller number of events have three clusters, which provides strong evidence for the existence of a high-momentum gluon radiated by the initial quark or antiquark ($e^+ + e^- \rightarrow q + \bar{q}$ + gluon). The results matched the detailed patterns predicted by the theory.

1979

**Standard Model Nobel Prize:** The Nobel Prize was awarded to Glashow, Salam, and Weinberg for their role in the development of the electroweak theory. This was four years before the observation of the $W^\pm$ and $Z^0$ bosons predicted by their theory, but after sufficient evidence for processes mediated by a $Z^0$ boson matching the predicted effects had been accumulated to convince physicists that the theory must be right.

1980

**Inflationary Cosmology:** Alan Guth proposed a new version of Big Bang theory in which the Universe begins with an initial period where a large cosmological constant causes a period of exponentially rapid expansion of space. Guth introduced this idea as a way to explain why we do not observe any magnetic monopoles, whereas grand unified particle theories predict their existence. The rapid expansion would dilute their presence to an unobservable low density. Guth recognized that the rapid inflation would also solve the horizon and flatness problems of standard Big Bang cosmology. His particular version of inflation does not work, but later variants have become the new standard cosmology.

1981

**CP Violation Test Proposal:** Ichiro Sanda and Ikaros Bigi proposed certain studies of B decays as tests of the CP violation predictions of the Standard Model. Most physicists at the time assumed these tests are not feasible.

1981–1982

**Inflationary Cosmology:** Andrei Linde, then in the Soviet Union and, independently, in the United States, Paul Steinhardt and his collaborators introduced new versions of the physics underlying an inflationary period (of exponentially rapid spatial expansion) in the early Universe. Most modern theories of inflation build on these versions, which predict patterns for the fluctuations in the cosmic background radiation since verified by experiment.

1983

**Observation of Z- and W-Bosons:** The $W^{\pm}$ and $Z^{0}$ intermediate bosons demanded by the electroweak theory were observed by experiments using the CERN synchrotron, which had been converted into a proton–antiproton collider by van der Meer and his team. The observations were in excellent agreement with the theory. The spin of the $W^{\pm}$ could be measured; it is $1\hbar$, as required by the Standard Model. The 1984 Nobel Prize was awarded to Carlo Rubbia and Simon van der Meer for this work.

## 1988

**Another Neutrino Anomaly:** The large underground neutrino detector Kamiokande (located in a mine in Kamioka, Japan) did not see the expected ratio of muon-type to electron-type neutrinos coming from cosmic ray events in the atmosphere.

## 1989

**Three and Only Three Light Neutrino Types:** Studies at CERN and SLAC of $Z^0$ boson production in $e^+ e^-$ collisions showed that $Z^0$ bosons decay to exactly three neutrino species, not more, strongly implying that there are only three generations of fundamental particles.

## 1989–1995

**Precision Testing of Standard Model:** Measurements of the mass and width of the $W^\pm$, at CERN and at Fermilab, together with the properties of the $Z^0$, provided further tests of the electroweak aspects of the Standard Model. Other tests, for example the energy dependence of the rate of production of multiple-particle clusters, added equally strong support for the strong interaction (QCD) aspect of the Standard Model.

## 1992

**Microwave Background Radiation:** The Cosmic Background Explorer (COBE) satellite was launched in 1989. Data from this experiment were taken over the next four years. It produced the first clean map of the tiny variations (at a few parts in a million) in the temperature of the black-body radiation observed in different regions of the sky. It also showed that the spectrum is indeed very precisely what thermal theory predicts. The temperature variations are the imprint of tiny density variations present at the time the Universe became transparent to light. All the current structure in the Universe must grow from gravitational effects due to these variations in density.

## 1994

**Red-Shift–Distance Relation from Supernovae:** Saul Perlmutter and his colleagues at Lawrence Berkeley Laboratory developed and implemented an automated method to search for supernovae. Supernovae found by this instrument could then be studied in detail by the Hubble Space Telescope. Over the next ten years this Supernova cosmology project and a similar effort known as the High $z$ Supernova search have provided an accurate and far-reaching red-shift–distance map.

## 1995

**Sixth Quark Discovered:** After eighteen years of searching at many accelerators, the CDF and DØ experiments at Fermilab discovered the top quark at the mass of about 175 GeV. No one expected so large a mass. In fact, to this day, no one understands any of the patterns of quark and lepton masses. The Standard Model can fit them but does not predict them.

## 1998

**Neutrino Masses:** Experiments with the larger neutrino detector super-Kamiokande show that the neutrino anomaly found in the earlier Kamiokande experiment is actually a difference between neutrinos coming from above and those coming from below. The latter travel a much longer path from production to detection, through the Earth. The expected ratio of electron type to muon type is found for the neutrinos coming from above, but there is a deficit of muon type coming from below. The best suggested explanation is that neutrinos have mass and that the missing muon-neutrinos have oscillated to become tau-neutrinos which are not detected in this experiment. The tide of physics opinion began to swing in favor of neutrino masses, but skepticism that this would also explain the solar neutrino effects remained strong.

1999

**Cosmology—Dark Energy:** Data from measurements of the relationship between red-shift and distance to very distant supernovae showed that the expansion of the Universe is not slowing down as expected, but rather is beginning to speed up. This requires a new feature of the theory, which has become known as dark energy. We do not have any real idea what it is, other than the fact that it is something that causes a very small value for Einstein's cosmological constant. Until this time, most cosmologists, unable to explain the smallness of this constant, had assumed it must be zero—hoping that would be more easily explained. Now we know that roughly 76% of the energy density in the Universe is in this quantity. You might ask why we call the quantity small if it is such a large fraction of the energy density—attempts to explain it from field theory give values that are more than 50 orders of magnitude bigger, and totally in conflict with the long-lived Universe that we find ourselves in.

2002

**Neutrinos Have Masses:** Combined results of all solar neutrino experiments, starting with those of Davis in 1968 and culminating in 2002 with the observations of the Sudbury Neutrino Observatory (SNO), experiment showed that the expected flux of neutrinos from the Sun does indeed arrive at the Earth, but that a significant fraction of them arrive not as the electron-neutrinos that were produced, but as muon- or tau-type neutrinos. This neutrino mixing effect proves that neutrinos must have mass. Since earlier experiments were sensitive to only electron-type neutrinos they saw a deficit. The Sudbury experiment was sensitive to all types. The results show that the neutrino states with definite mass are mixtures of the states produced in association with a given charged lepton type. The anomalies in atmospheric neutrinos (see 1988, 1998) can also be explained in a similar way; they show mixing between muon- and tau-type neutrinos. Finally, the majority of the physics community was convinced—there was no longer any doubt that neutrinos have very small masses.

1999–TODAY

**CP Violation:** Two B-factory experiments, BaBar at Stanford Linear Accelerator Center in California, and Belle at the KEK lab in Japan, began running in 1998. By now they have collected data on millions of B-meson decays. These experiments have observed CP violation in B-meson decays as predicted by the Standard Model, and are continuing to work to provide more sensitive tests of the predictions of the Standard Model for this and related effects.


2003–2004

**Concordance Cosmology:** Starting with the COBE satellite, then the DASI telescope at the South Pole, and the Boomerang and Maxima balloon experiments, and finally the WMAP satellite launched in 2001, data gave enough detail on the distribution of fluctuations in the temperature of the cosmic background radiation to test cosmological models. A concordance model of the parameters describing the evolution of the Universe emerged. It fitted the evidence obtained in multiple independent studies, including data on cosmic background radiation, on red-shift–distance relationships for supernovae, on dark matter from studies of galaxies, and on visible matter densities from deep sky galaxy surveys and from simulations of the evolution of structure in the Universe. The results conformed to the expectations of inflationary Big Bang cosmology in most respects but also required a cosmological constant or some physical process that would give an effective cosmological constant. The general term for this latter quantity is dark energy. As yet, no one has a clue what it is, though there are many suggestions. The Physics Nobel Prize in 2006 was awarded to John Mather and George Smoot for their leadership of the COBE Satellite experiments.


2006

**Summary:** Physicists and astrophysicists have two strong Standard Models, one for particle physics, one for cosmology, but major puzzles remain. Is there a Higgs boson? (Is the Standard Model mechanism for fundamental particle masses correct?) What more massive particles exist? Do they fit any of the predicted patterns? Are protons ultimately unstable particles? What particles form the dark matter in the Universe? What is dark energy?

What features of the theory account for the matter–antimatter imbalance of the present-day Universe? If the past is any guide, the answers to some of these questions will require major additions and changes to our "standard" models. We are already speculating on what these changes might be, and how one would test the new ideas.

# *INDEX*

*When multiple page numbers occur for an item, the page number for the main discussion of that item is in boldface type.*